EAI/Springer Innovations in Communication and Computing

Series Editor

Imrich Chlamtac, European Alliance for Innovation, Ghent, Belgium

The impact of information technologies is creating a new world yet not fully understood. The extent and speed of economic, life style and social changes already perceived in everyday life is hard to estimate without understanding the technological driving forces behind it. This series presents contributed volumes featuring the latest research and development in the various information engineering technologies that play a key role in this process. The range of topics, focusing primarily on communications and computing engineering include, but are not limited to, wireless networks; mobile communication; design and learning; gaming; interaction; e-health and pervasive healthcare; energy management; smart grids; internet of things; cognitive radio networks; computation; cloud computing; ubiquitous connectivity, and in mode general smart living, smart cities, Internet of Things and more. The series publishes a combination of expanded papers selected from hosted and sponsored European Alliance for Innovation (EAI) conferences that present cutting edge, global research as well as provide new perspectives on traditional related engineering fields. This content, complemented with open calls for contribution of book titles and individual chapters, together maintain Springer's and EAI's high standards of academic excellence. The audience for the books consists of researchers, industry professionals, advanced level students as well as practitioners in related fields of activity include information and communication specialists, security experts, economists, urban planners, doctors, and in general representatives in all those walks of life affected ad contributing to the information revolution.

Indexing: This series is indexed in Scopus, Ei Compendex, and zbMATH.

About EAI - EAI is a grassroots member organization initiated through cooperation between businesses, public, private and government organizations to address the global challenges of Europe's future competitiveness and link the European Research community with its counterparts around the globe. EAI reaches out to hundreds of thousands of individual subscribers on all continents and collaborates with an institutional member base including Fortune 500 companies, government organizations, and educational institutions, provide a free research and innovation platform. Through its open free membership model EAI promotes a new research and innovation culture based on collaboration, connectivity and recognition of excellence by community.

Anveshkumar Nella • Anirban Bhowmick
Chandan Kumar • Maheswar Rajagopal
Editors

Energy Harvesting Trends for Low Power Compact Electronic Devices

Springer

EAI
RESEARCH MEETS INNOVATION

Editors
Anveshkumar Nella
School of EEE, VIT Bhopal University
Kothri Kalan, Sehore
Madhya Pradesh, India

Anirban Bhowmick
School of EEE, VIT Bhopal University
Kothri Kalan, Sehore, Madhya Pradesh, India

Chandan Kumar
Department of Biomedical Engineering
Central University of Rajasthan
Ajmer, Rajasthan, India

Maheswar Rajagopal
Department of ECE, Centre for IoT
and AI (CITI)
KPR Institute of Engineering and Technology
Coimbatore, Tamil Nadu, India

ISSN 2522-8595 ISSN 2522-8609 (electronic)
EAI/Springer Innovations in Communication and Computing
ISBN 978-3-031-35967-5 ISBN 978-3-031-35965-1 (eBook)
https://doi.org/10.1007/978-3-031-35965-1

This Springer imprint is published by the registered company Springer Nature Switzerland AG
The registered company address is: Gewerbestrasse 11, 6330 Cham, Switzerland

Editor's Note

In this growing digital world, the development of low-power electronics and wireless sensors has led to a wealth of possibilities. Due to the wireless nature of these systems, the devices need their power source. The devices are usually powered by a battery with a limited power supply. There is however a cost and size increase associated with batteries as well as a burden associated with the replacement of batteries or the charging of the device in question. To meet the energy demands of these wireless communication systems, a growing amount of work is being done to find new, sustainable energy sources. There are several ways of addressing this challenge, including (i) using a locally available source of energy, (ii) using a wireless power generation source, and (iii) harvesting energy from the environment efficiently. Energy harvesting or energy extraction refers to the process of recovering energy from alternative sources in the environment. Typically, this involves the recovery of unused or excess energy, which may be a by-product of a natural phenomenon or an automated process and is therefore considered free energy. This energy can be used to make wireless and portable electronic devices completely self-sufficient, eliminating the need for battery maintenance or at least providing a backup in the event of an emergency or failure of the main power system. Examples of potential energy sources include electromagnetic energy in the form of solar energy, infrared energy, or RF; mechanical energy in the form of vibration, voltage, and stress; thermal energy from furnaces, combustion engines, and other heat sources; human energy from various metabolic processes and functions inside or outside the human body; sound energy; etc. The energy from these sources can be detected by various sensors and converted into usable energy, i.e., electrical voltages and currents. This can be harvested, stored, and further processed according to the requirements of wearable electronics and wireless sensor networks.

Kothri Kalan, Bhopal, India — Anveshkumar Nella
Kothri Kalan, Bhopal, India — Anirban Bhowmick
Ajmer, Rajasthan, India — Chandan Kumar
Coimbatore, Tamil Nadu, India — Maheswar Rajagopal

Preface

A recent increase in the use of low-power electronic devices is a testament to their incredible capabilities and efficiency. They use low power supplied by batteries or other power sources. Additional power sources for these devices lead to an increase in the size of the system and higher costs. Therefore, unconventional sources of energy and waste materials can be used to operate various electronic devices with low power consumption. Energy harvesting is the process of extracting energy from the environment and converting it into usable electrical energy. Like other applied sciences, energy harvesting has been formulated and processed in various ways, depending on the energy source used.

This textbook mainly focuses on the numerous energy harvesting techniques and their system implementation to meet the energy requirements in compact electronic devices. In addition to mobile phones and portable devices, these devices are also used for bio-medical services, defense and military systems, agriculture needs, mechanical system aspects, sensor networks, automobiles, food, home appliances, and industry needs. This book covers the latest energy harvesting methods such as acoustics, heat, artificial light, fluid flow, mechanical vibration, EM energy, RF energy in 4G and 5G, piezoelectric, electrostatic, photovoltaic, thermoelectric, hybrid, ultrasonic, and the popular wind- and solar-based energy harvesting methods. The energy extracted from the above sources can be used for the Internet of Things, wireless sensor networks, biomedical devices, wearable devices, smart devices, bioelectronic systems, implantable medical and health monitoring devices, body area networks, security devices, etc., and would therefore be useful to industry persons and researchers.

Kothri Kalan, Bhopal, India — Anveshkumar Nella
Kothri Kalan, Bhopal, India — Anirban Bhowmick
Ajmer, Rajasthan, India — Chandan Kumar
Coimbatore, Tamil Nadu, India — Maheswar Rajagopal

Acknowledgements

The editors would like to express their gratitude to God for his uncountable kindness in making this book a resounding success. Additionally, the editors would like to express their gratitude to their families, friends and colleagues who continually encouraged them to complete the project on time. Aside from that, they wish each other the best of luck with the book's success.

The editors are also thankful to the Springer Editorial Team, who really helped them in every aspect of the preparation of this book. Sincere thanks to Eliska Vlckova, Managing Editor, European Alliance for Innovation (EAI), for her help in approving the book proposal and providing us with the chance to edit the book under EAI, Springer. As a remarkable and sensational contributor, she played an essential part in supporting, responding, providing guidance, and taking care of the literature and contributions.

The editors are also highly thankful to the Management and Institute officials of VIT Bhopal University, Sehore, Madhya Pradesh, India; Central University of Rajasthan, Ajmer, Rajasthan, India; and KPR Institute of Engineering and Technology, Coimbatore, Tamil Nadu, India for their massive support and encouragement. Without their assistance, this book would not have been possible for getting published.

The editors would like to thank all contributing authors for their excellent manuscripts, reviewers for providing valuable suggestions and comments on improving the chapters at all times, and everyone behind the successful completion of this book.

Contents

Material and Component Selection for Efficient Energy Harvesting

N V R Vikram Gelli

1 Introduction

Technologies and materials for energy harvesting to harvest energy from the environment needed to be developed to power various smart and portable devices. These devices need batteries that need to be replaced or recharged over a period. While it is convenient to replace or recharge them in a few cases, it will be a difficult task to move a few of them to a different place for charging. In some cases, the batteries being used may be hazardous and bulky [1–4]. Hence, they need an alternate energy source for better sustainability. The energy from the surrounding environment is capable of powering the devices over the long run and eliminating the need for batteries in some cases. The energy that can be harvested is available in the environment in various forms such as RF, heat, vibrations, sound fluid flow, and light, to name a few. However, the energy that can be harvested from these sources is very limited (Table 1) owing to the nature of the availability. There are multiple studies to harvest the energy from these sources, which use different components/materials for harvesting the energy and storing/using it. Proper components/materials must be used for harnessing and supplying the minimal amount of energy available from the abovementioned sources to power up the electronic devices continuously for their long-term usage. A simplistic energy harvesting system (Fig. 1) contains:

1. Ambient energy: Various forms of energy in the environment.
2. Energy harvester: The device that harnesses the energy and converts it into useful electrical form.

N V R V. Gelli (✉)
Department of ECE, Vignan's Foundation for Science, Technology, and Research, Guntur, Andhra Pradesh, India

A. Nella et al. (eds.), *Energy Harvesting Trends for Low Power Compact Electronic Devices*, EAI/Springer Innovations in Communication and Computing,
https://doi.org/10.1007/978-3-031-35965-1_1

Table 1 Energy harvester power densities

Solar cells	21 mΩ/cm^2
Piezoelectric [5]	3.53 mW/cm^3
Thermoelectric [6]	1.4 mW/cm^2
RF [7]	84 nW/cm^2
Vibration	116uW/cm^3
Acoustic noise (100 dB)	960nW/cm^3

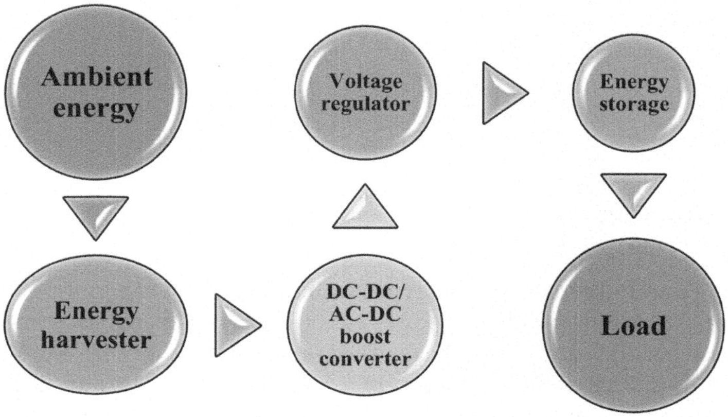

Fig. 1 Illustration of energy harvesting system

3. DC-DC/AC-DC boost converter: Boosts the electricity generated by the harvester to a reasonable level to be stored/used.
4. Voltage regulator: To supply a fixed voltage to the energy storage unit/load.
5. Energy storage: To store the electricity supplied by the harvester for future use.
6. Load: The device uses the energy from the harvester.

In this chapter, we discuss how various forms of energy can be harvested using a selective component/material to improve the performance of the whole harvesting system. More emphasis in the chapter will be on the components used for harvesting and the efficiency of the systems. It will also cover the various materials that are used for the harvester. In the end, a summary of the various components/materials used for harvesting will be presented.

2 Component/Material Selection

As discussed in the introduction, one needs to understand what are the different components used for energy harvesting along with the materials used for the harvesting devices. The various energy sources that are under discussion are RF energy, acoustic energy, piezo-energy, mechanical energy, fluid energy, light energy, and thermal energy.

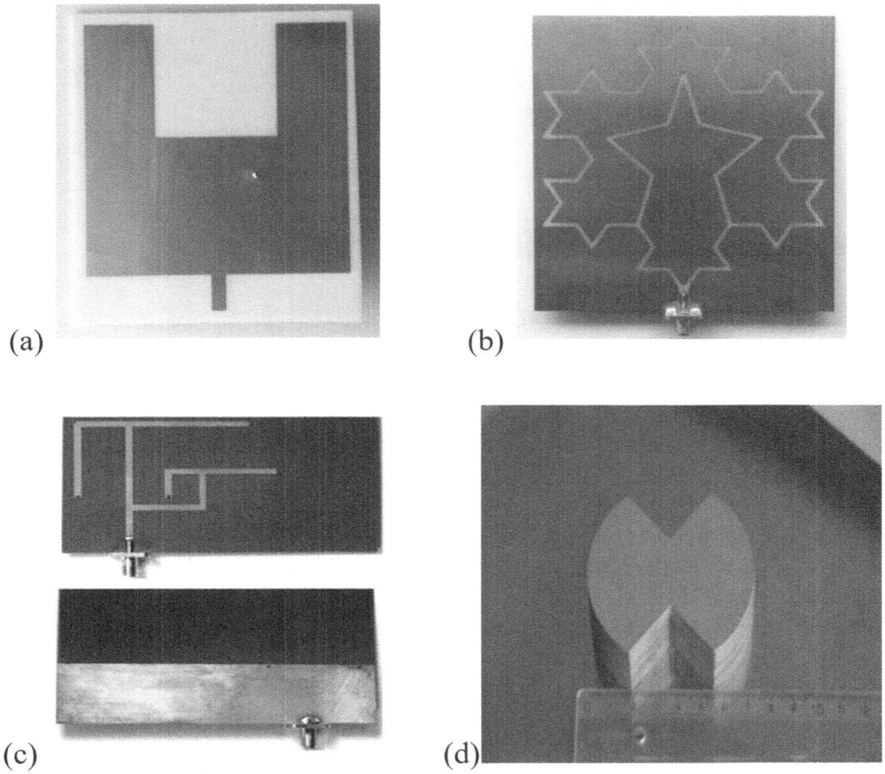

Fig. 2 Manufactured multiband patch antenna (a), dual-band Koch fractal antenna (b), dual-band IFA (c), and wideband bow-tie DRA (d). (Reproduced from [8])

2.1 RF Energy

RF energy harvesters harvest energy using an antenna, RF-DC converter, and a voltage multiplier. The common materials used for antenna fabrication are gold and copper. Hence, the discussion here will be more on the design of the antenna. The antenna and its design play a crucial role in determining the energy harvested. Four different structures, viz., patch, slot, dielectric resonator, and modified inverted F, were compared for their performance at different frequencies from 0.8 to 2.6 GHz [8]. Each of the antennas gave peak efficiencies at different frequencies: patch antenna, 2.45 GHz; slot antenna, 1 GHz; dielectric resonator antenna, 1.9 GHz; and inverted F antenna, 1.7 GHz (Fig. 2).

The sensitivity and gain of the antenna were found to be increasing with the increasing number of patches in the antenna [9]. However, the gain was more for the 2 × 2 array patch compared to the 4 × 1 array patch [10]. For very low-input RF powers, of the order of 1 μW/cm^2, a single-diode rectifier could be employed for its

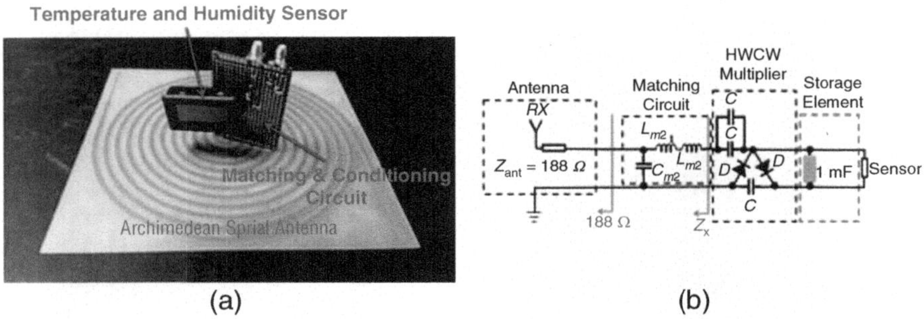

Fig. 3 Full RF harvesting system including antenna and matching and conditioning circuit (**a**) and its circuit diagram. (Reproduced from [13])

better efficiency of RF-DC conversion [11]. Rectenna – antenna integrated with rectifier on the substrate – could minimize the losses and provide higher efficiencies when RF power is lower [12].

A half-wave Cockcroft-Walton circuit and a capacitor coupled with an Archimedean spiral antenna harvested the majority of power (82%) in the LTE and GSM bands of 800 and 900 MHz, and the rest was distributed over 1800, 2100, and 2600 MHz and WIFI, FM, WiMAX, and DTT. The efficiency of the system is about 30% with the voltage across the capacitor of 1.25 V to switch on a small temperature and humidity sensor (Fig. 3) [13].

An array of long spiral antennae with half-wave rectifier produced energy with a circuit efficiency of 20% [14]. A linear slot tapered antenna with a differential rectifier had 13% efficiency for 1.85 GHz frequency [15]. Concentric square patched antenna with full-wave rectifier at 2.4/5.5 GHz frequency had 36%/5% efficiency [16]. Half-wave Cockcroft-Walton circuit paired with a slotted patch antenna could achieve 68% efficiency for 2.45 GHz frequency [17].

2.2 *Acoustic Energy*

Harvesting of acoustic energy is done by first collecting the ambient sound wave, then amplifying it by a resonator or acoustic metamaterial, and converting it to electric energy. At the last, it needs to be rectified and regulated for storage or to power electronics. Sound energy is another good candidate for energy harvesting that has not been used properly [18].

An organic thin film-based triboelectric generator has been employed to convert the acoustic noise from the vibration source to electricity. It uses PTFE/Cu bimorph diaphragm with porous electrodes. PTFE nanowires act as triboelectrifiers. The sensitivity of the device is 9.5 V/Pa. A peak power density of 60 mW/m^2 has been achieved, which can be used to power LEDs or the self-powered microphone [19]. Other methods of acoustic energy harvesting include piezoelectricity and

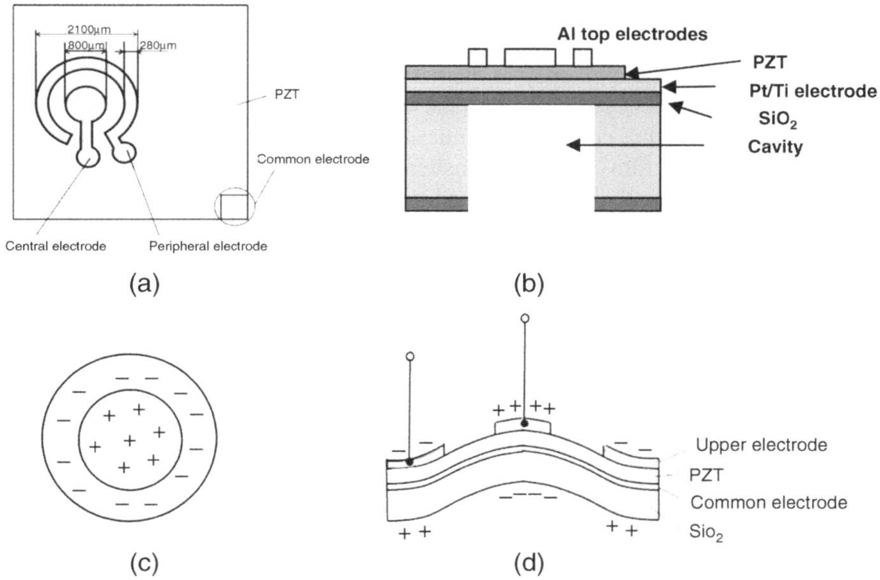

Fig. 4 Structure of PZT energy harvester; (**a**) top view, (**b**) schematic cross-section, (**c**) expected charge distribution, and (**d**) concept. (Reproduced from [21])

electromagnetism. A Helmholtz resonator is one of the best elements for acoustic energy harvesting [20]. A circular truncated cone shape cavity resonator provided better stiffness compared to a cylindrical-shaped cavity, thus improving the pressure gain. Under a sound pressure level (SPL) of 130 DB, it produced a maximum power of 214.23 μW. The bandwidth of this resonator ranged from 1453 to 1542 Hz, 1710 to 1780 Hz, and 1848 to 1915 Hz.

A piezoelectric capacitor was formed with a layered structure of Al/PZT/Pt/Ti/SiO_2 with thicknesses of 0. 1/1/0.1/0. 1/1.5 μm, respectively, above a 2 mm diameter cavity in a silicon wafer of 300 μm thickness. As the frequency of the acoustic signal increases, the output voltage increases. A maximum power density of 20.7 $\mu W/m^2$ was achieved with a resistance of 25 ohms [21] (Fig. 4).

2.3 *Piezo-Energy*

A gradient of electric potential will be developed across piezoelectric materials when mechanical strain is applied to them. The electric field is proportionate to the stress resulting from the strain. The mechanical strain which is resulting from vibrations can be transformed into alternating current for harvesting energy from vibration using piezoelectric materials. The vibrating portion is generally a piezoelectric cantilever beam. To reduce the resonant frequency, a proof mass is attached to it. Low-stiffness cantilevers are needed to capture the vibrations with low frequencies. Similarly, two layers of piezoelectric material can be used as a bimorph structure to improve the efficiency of the harvester by providing a larger strain at a lower resonance frequency compared to the structures with a single layer [22].

Piezoelectric materials comprise ceramics, single crystals, relaxor-type ferroelectrics polymers, thin films, and composites. The majority of the piezo-energy harvesting materials used are ceramics and polymers [23–26]. Single-crystal piezoelectric materials like lithium tantalate, quartz, and lithium niobate have also been explored for their harvesting properties [27]. Though the maximum electricity generated by lithium niobate is less than lead zirconia titanate (PZT) ceramic, its conversion efficiency (78%) was more compared to PZT (68%). Perovskite structured ceramic piezoelectric materials lead zirconate-lead titanate (PbZrxTi1 − xO3)-PZT, barium titanate (BaTiO3), and lead titanate (PbTiO3, PCT) can be easily tailored for a multitude of applications [24, 25].

An off-the-shelf PZT harvester rectifier board generated 3.072 mW at a frequency of 53 Hz [28]. An increase in the resonant frequency resulted in decreased power output. To transmit a signal to a transceiver over a short distance, a PMN-PT piezoelectric disk was used to supply 160 μJ energy at 2.5 V based on a single button pressing [29].

PVDF and PZT-based integrated piezo-fiber were used to generate power from inside the tire during the travel. It produced minimum energy of 200 μJ per cycle when a load of 200 kgf is applied. The maximum energy of 380 μJ is produced when the load is 450 kgf. The velocity during the travel was 60 km/h. This energy would be sufficient to power a node of a wireless sensor network [30].

2.4 Mechanical Energy

Mechanical energy sources can be vibration, intermittent, and steady state. The vibration source is the best among them. Fluid flow like wind, water flow, and air currents are some examples of steady-state sources.

Vibration energy from an air compressor in the cold storage facility was harvested using an electromagnetic vibration harvester. At frequencies of 49.3–49.7 Hz and acceleration of 25–48 mg (milli g), the system in combination with the bridge rectifier was able to generate 1.56 mW [31].

Perylene cantilevers with a metal line in combination with magnets attached to a diaphragm for capturing ambient vibration were used as electromagnetic generators [32]. This setup can harvest energy from environmental vibrations at low frequencies as the structure acts as a frequency upconverter to generate higher-frequency signals. From environmental vibrations with frequencies ranging from 70 to 150 Hz, a generator of 0.15 cm^3 size could generate a power of 0.25 mW and voltage of 0.57 mV through harvesting. As the number of coils is increased, voltage levels can be improved.

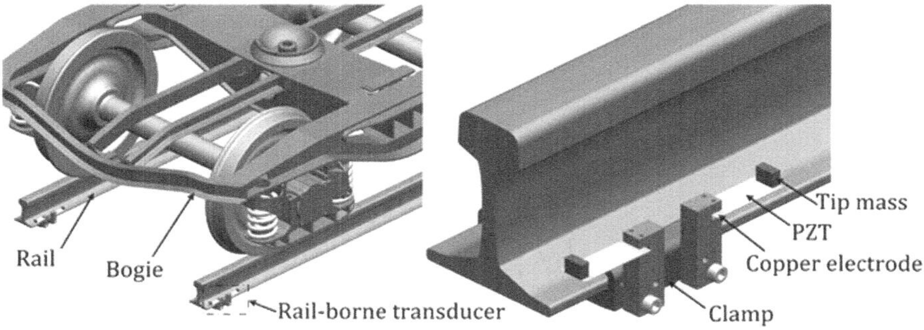

Fig. 5 (left) Illustration of bogie-rail-harvester scales; (right) enlarged view of clamped cantilevered piezoelectric beam configuration. (Reproduced from [33])

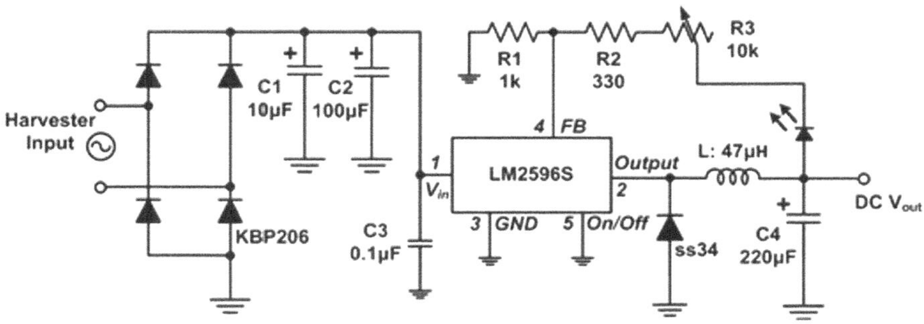

Fig. 6 (**a**) Schematic and (**b**) PCB prototype of bridge rectifier/DC-DC buck converter circuit. (Reproduced from [33])

Seismic energy from rails was harvested using a PZT film-based cantilever (Fig. 5). It could generate 4.9 mW of maximum power at 22.1 V with a load impedance of 100 ohms [33], while the rail acceleration was 5 g. A peak voltage of 32 V was achieved at a 7 Hz frequency. AC voltage to DC voltage conversion was carried out using a bridge rectifier and a DC-DC buck converter (Fig. 6).

2.5 Fluid Energy

Air and water flows are a type of fluid energy source for harvesting. They can be either natural or artificial and could be from rivers, wind, ventilation ducts, or water pipes. To harvest the energy from the environment to power portable electronics, regular harvesting mechanisms that use turbines, electromagnetic induction, and piezoelectricity could be employed to harvest energy from these sources. A flow channel with a pulse pump along with an electromagnetic diaphragm was used to harvest energy from tap water flowing through a pulse pump [22]. Power 0.4 μW

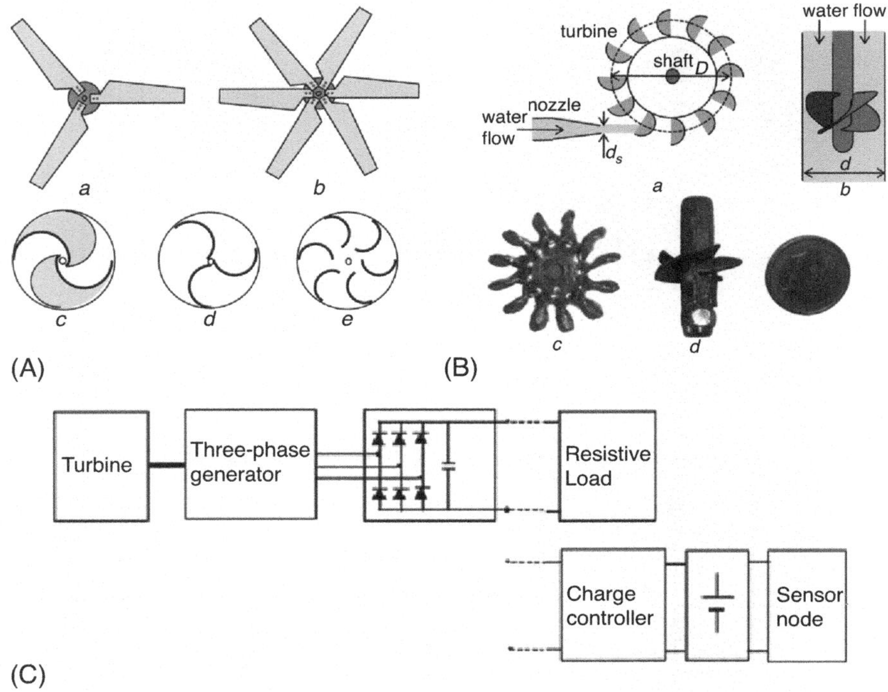

Fig. 7 HAWT and VAWT wind turbines (**a**), hydro Turbines (**b**), and harvesting circuit layout (**c**). (Reproduced from [35])

was generated at a pulse frequency of 30 Hz and water pressure of 254 Pa. A similar energy harvesting system with a piezoelectric harvester on a PDMS diaphragm was developed to harvest from oscillatory fluid pipelines [34]. The piezoelectric film harvested electricity from these oscillations due to the fluid flow through the pipes. At an excitation pressure of 1.196 kPa and 26 Hz frequency, an open circuit voltage of 2.2 V at a power of 0.2 μW was generated. It can be used to power small and portable electrical appliances that require such low power.

Mini turbines to harvest energy from water and wind flow have been designed and tested. The power generated in the range of 41 mW to 1256 mW was sufficient to power nodes of wireless sensor networks [35]. Horizontal axis wind turbines (HAWT) with three and six blades (Fig. 7a, b) and vertical axis wind turbines (VAWT) with two blades (dia, 4.2 cm), three blades (dia, 3.2 cm), and six blades (dia, 2.4 cm) (Fig. 7c, d, e) were used to harvest the wind energy. Three-blade HAWT had higher efficiency at larger water flow, and six-blade HAWT gave better efficiency at medium water flow. Though the efficiency of VAWT was small compared to that of HAWT, their efficiencies were better at medium water flow.

Mini Pelton turbine (with a radius of 1.6 cm, Fig. 7a, c) and a mini propeller turbine (of diameter 2.7 cm, Fig. 7b, d) in combination with a coil and permanent magnet were used to harvest electricity from water flow [35]. The power generated ranged from 42 to 1125 mW. Pelton-based designs produced higher power with better efficiencies compared to propeller-based turbines. Pelton turbines were driven by water flow from a nozzle with a diameter varying from 2.2 to 3.2 mm. With the increased diameter of the nozzle, the efficiency of harvesting also increased. As the water head height increased, the efficiencies of both the Pelton and propeller-based designs improved. However, four-bladed propellers showed decreased efficiencies with increased head height. To harvest the energy, the turbines are connected to a three-phase generator, and six-diode-capacitor rectifier (Fig. 7c) circuit was used. The output of the circuit was DC voltage which was used to charge battery via charge controller. The output of the battery was used to power a small sensor node.

2.6 Light Energy

Light energy harvesting can be done from either ambient light indoors or outdoors from solar energy. Ambient light is a typical example [36–44]. Light is in a certain portion of the electromagnetic spectrum. The wavelength of visible light is in the range of 400–700 nm (nm), between the ultraviolet rays and the infrared rays. This translates to a frequency range of approximately 430–750 terahertz (THz).

A light energy harvester can usually eliminate the need for a battery [36]. Light intensity as low as 200 Lux can produce power at room temperature under fluorescent lamps. Maximum power per square inch increased linearly with incident illumination. Ultracapacitors and batteries can be used to store energy. Ultracapacitors do not need any circuitry for charging, while batteries need special circuits. Non-rechargeable battery + forward-biased diode combinedly in parallel with a PV (photovoltaic) cell can be used to power the devices without any interruption. When enough light intensity is available, PV cell powers the device, and in the absence of light, the battery supplies the power. Similarly, PV cell + forward-biased diode combinedly in parallel to a rechargeable battery, with a battery charging circuit (Fig. 8), can power the devices [37, 38]. The rechargeable battery can be replaced with an ultracapacitor. A DC-DC boost converter and a buck-boost converter (Fig. 8) can be used to convert and control the voltage generated from the PV and to charge the battery. This setup is very useful for powering wearable sensors which provide standard batteries, and replacement will be a difficult task. A small PV cell of 16 mm^2 has an efficiency of about 77–89% with 1 mW output [38]. This is sufficient to power small sensors in the indoor environment.

Energy harvested from PV cells, under indoor lighting conditions, through a novel design with an MPPT (maximum power point tracking) circuit (Fig. 9) was stored in a supercapacitor to accumulate power and deliver for a long period [39]. A DC-DC step-up converter with reduced power loss and a start-up signal circuit was used for more efficiency. The MPP ("maximum power point") tracking sensitivity can be adjusted by adjusting the hysteresis of comparator 1 which was used as a control unit. When the harvester is made to work at MPP, the circuit can power the

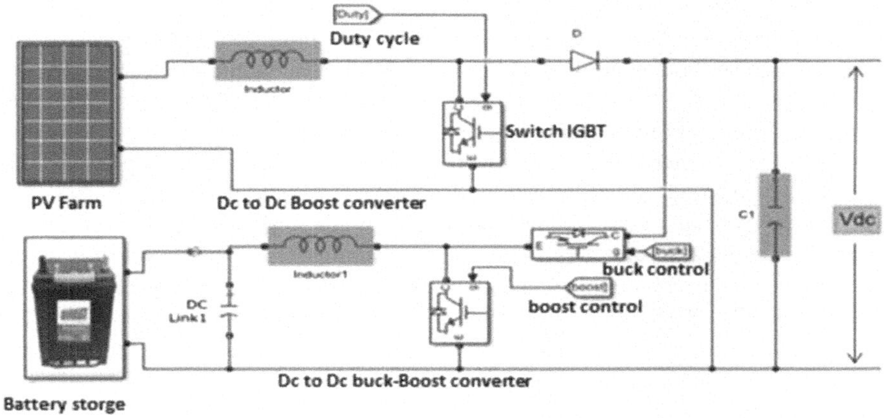

Fig. 8 The boost of PV and the buck-boost converter of the battery. (Reproduced from [37])

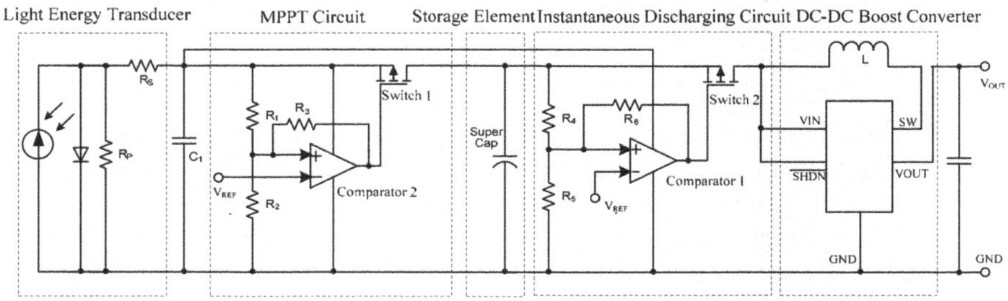

Fig. 9 The block schematic of the proposed power management system. (Reproduced from [39])

sensor node by harvesting indoor faint light energy. The energy thus harvested was stored in a supercapacitor. A DC-DC boost converter with a low inductance of 10 mH and a supervisor setup was implemented to reduce the power loss in the inductor and reduce the unnecessary power consumption when supercapacitor voltage is low. This could drive the wireless-smart humidity and temperature sensor that requires a power of 105 mW/cycle when the input power from indoor light was at 74 72.74 μW [39].

Solar cells placed on low-power consuming devices or smartphones (Fig. 10) were used along with a mechanism similar to that shown in Fig. 9, harvesting energy from ambient light to power those devices [40]. In this case, the internal batteries of the devices were used for storage. To enhance harvesting from light energy, a novel mechanism of solar concentrators was used. A monocrystalline PV cell (KXOB22–04 × 3 L (IXYS)) of 154 mm^2 area was used in this study. At 25 °C and λ = 1000 W/m^2, the I_{SC} (short circuit current) is 15 mA and V_{OC} is 1.89 V for this cell and the efficiency is 22%. The time taken for charging to 10% of the battery at 2500 Lux (office environment) was about 8 h with four PV cells. The

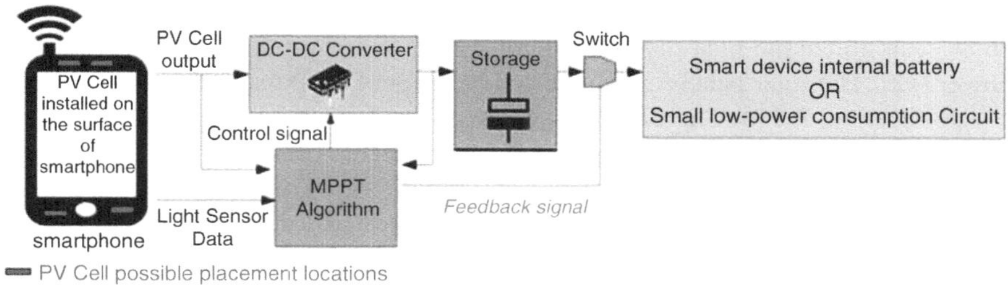

Fig. 10 Complete system setup for harvesting ambient light energy to power smartphone. (Reproduced from [40])

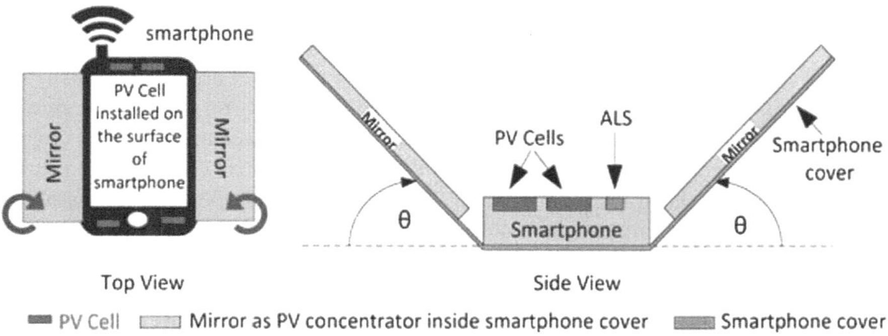

Fig. 11 Solar concentrators using mirrors to enhance the incident light. (Rproduced from [40])

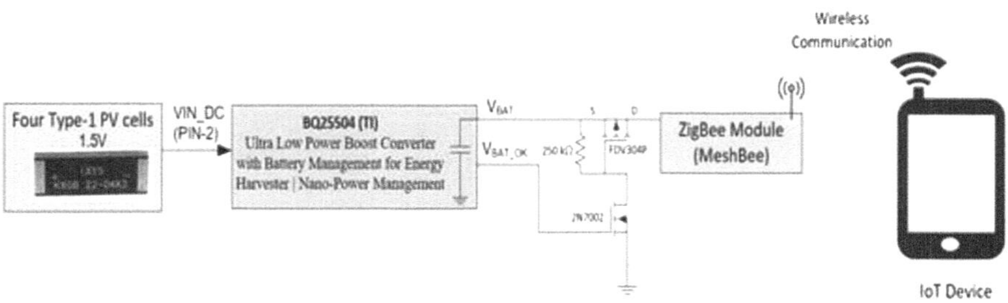

Fig. 12 Powering Zigbee module for testing the harvested energy. (Reproduced from [40])

enhancement of the light intensity at two angles 71° and 90° were 1.43 and 1.85, respectively. Another setup with type 1 PV cells was implemented to test by powering Zigbee communication effectively (Figs. 11 and 12).

IoT devices and smartphones could be charged through a system based on lasers [41]. In this system, the power was transmitted across the room without exposing humans to laser and provided an end-to-end wireless power solution. A power of 2 W was delivered to the receiver of the table-top form factor and a smartphone. Even before a human enters the laser beam area, up to a speed of 44 m/s, the laser source can be turned off for safety.

Solar cells made with improved triple junction (ITJ) with efficiency rated at 26.8% can utilize the majority of the light spectrum and produce extreme possible power [42]. The solar panels are connected in series via bypass diodes to reduce the hotspot heating. A blocking diode can be used to reduce the damage from the dark current flowing out of the battery during no sunlight. Changing the PV voltage module periodically improved the power output, which was tracked by MPPT (maximum power point tracker). The energy harvested using the solar cells was stored in a battery using a low-voltage monolithic step-up converter IC SPV1040. This energy harvested can thus be used at any time to power up a railway track monitoring system for necessary corrective action.

A bendable PV panel (60 mm × 72 mm) was used to power electronics and body sensors for distributed biometric monitoring [43]. A bendable amorphous PV panel along with MPPT and supercapacitor combinedly was referred to as FEH (flexible energy harvester). The PV panel produced a peak power of about 600 μW at an intensity of 1000 Lux. The peak power harvested by the panel at a light intensity of 320 lux (indoor) was 77 μW. Even when the panel was bent (up to 30°), the power harvested had dropped only by about 2 μW. A flexible power management circuit (FPMC) was used to reduce the bulkiness. There was a loss of 11.5% in the power generated due to the implementation of FOCV (fractional open-circuit voltage)-based MPPT. The maximum electrical loading was 36.39 μW and a leftover power of 7.71 μW could be stored in a supercapacitor. This FEH can be used for various biomedical applications to reduce the need for a battery.

A combination of piezoelectric and photovoltaic flexible strips was used to harvest the energy from solar and wind together [44]. It used an inverter flag that is a combination of flexible strips of photovoltaic and flexible strips of piezoelectric together (Fig. 13). The power generated by the system when the range of speed of the wind was 0–26 m/s and the light that was incident was 1.8 kLux was 3 mW to 4 mW.

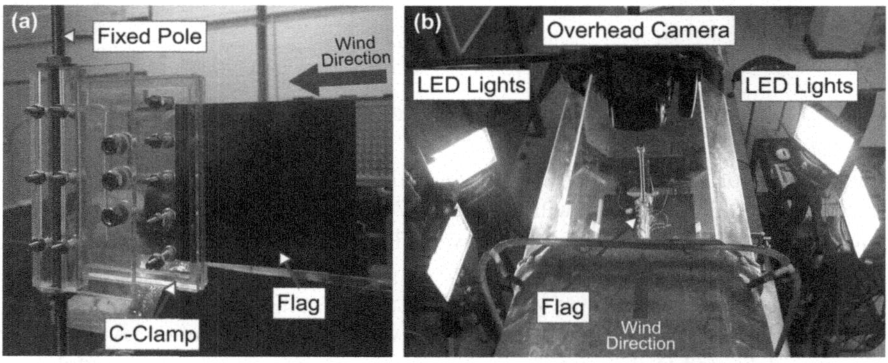

Fig. 13 (**a**) Close-up of clamping method and (**b**) wind tunnel setup. (Reproduced from [44])

2.7 Thermal Energy

Thermal energy is one of those which is being wasted in many industrial components, human bodies, and household devices. It can be harvested to generate electricity that can power small sensor-based devices [1, 4, 31, 45–51]. A TEG (thermoelectric generator) is a device that is used to convert heat energy into electricity based on the Seebeck effect. The Seebeck effect, discovered by Thomas Seebeck in 1821, is the generation of voltage across two junctions of dissimilar materials maintained at different temperatures (Fig. 14). The amount of energy harvested increases with increasing temperature difference (T_d). To maintain the energy of harvesting constant, one needs to remove the heat from the cold surface by using a heat sink or a fan (Fig. 15).

Fig. 14 Thermoelectric generator working principle. (From [46])

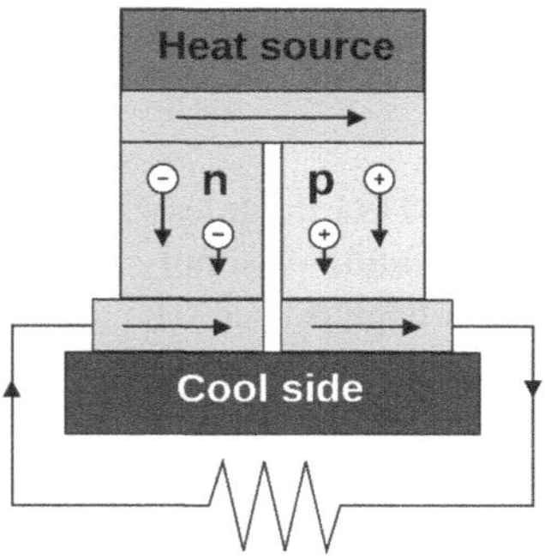

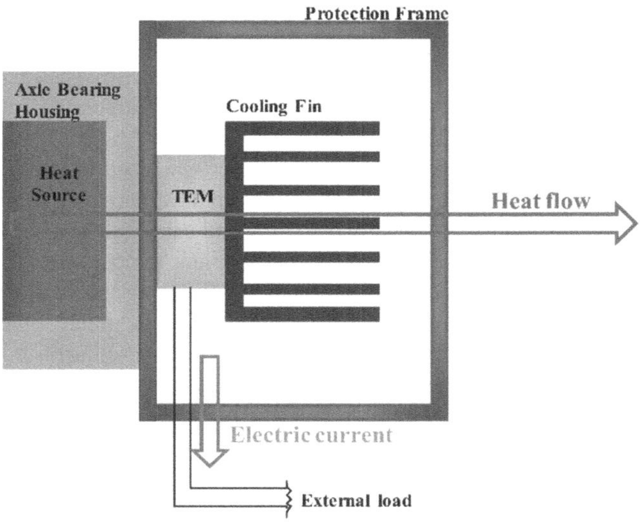

Fig. 15 System architecture of thermoelectric energy harvesting system. (Reproduced from [50])

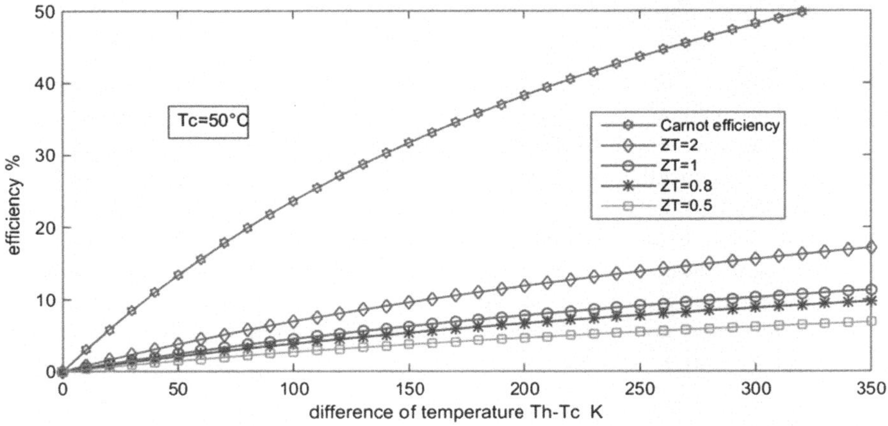

Fig. 16 Typical values of TE efficiency for different values of ZT. (Reproduced from [45])

The TEG modules can be used to harvest energy from a heat source and store it for the future or supply power directly to small electronic devices. It generates very low voltage/power. The amount of voltage thus the efficiency of harvesting by the TE (thermoelectric) module is dependent on the T_d between the junctions and the product of average temperature (T) by a factor of merit (Z) – ZT (Fig. 16) [45]. The efficiency of harvesting increases with the T_d and the factor of merit as well. The majority of the materials used for TEG are II–IV group materials as they offer low-cost and zero maintenance. Various material combinations that can be used for TEG are listed in Table 2. They can operate up to a maximum T_d of 750 K and an operating temperature of 800 °C.

Thermal energy can be harvested from the human body through clothing containing TEG modules to power wearable sensors/low-power electronics [47, 48]. A series of 14 very thin TEG modules were integrated into the shirt along with PV cells to harvest energy both from heat and light [47] to power an EEG device. A fully integrated DC-DC upconverter was used to charge the NiMh rechargeable battery from the energy harvested from TEG modules (Fig. 17). Maximum power of 8 mW and V_{OC} of 5 V were harvested using the TEG modules when the person wearing the shirt was walking, at an ambient temperature of 16 °C. The power and the V_{OC} were reduced to 3 mW and 3.2 V during the standstill position. As the ambient temperature reduced, the output fell to a much lower value. A microcontroller was used to monitor the output of TEG and control the number of stages in the DC-DC upconverter charge pump. A flex mechanism was developed to make the electronics module compatible with the cloths (Fig. 18).

The temperature difference between the soil and the bottom of the rail could be used to generate electricity using TEG modules [49]. Electricity can be produced at the available difference up to 30 °C. A DC-DC booster was used to store the harvested energy in Li batteries. At a T_d of 8 °C, the power was 5.8 mW and it

Table 2 List of TE modules and their properties

Manufacturer	Materials	Temperature difference ΔT	Power weight	Status	Maximum temperature	Information, outlook
HiZ, Thermonamic, Lairdtech, Marlow, Komatsu etc.	$Bi_2\,Te_3$	300 K	20 W 115 g	€40–€100	300 °C	Scarce (rare earth), toxicity
Evident thermoelectric	Half-Heusler	500 K	15 W	Coming soon	600 °C	Environmentally-friendly, low cost, availability of raw materials
Shanghai Institute of Ceramics	Skutterudites	510 K	25 W	Coming soon	600 °C	Environmentally-friendly, low cost, availability of raw materials
TEGMA	Skutterudites			s still in development		Environmentally-friendly, low cost, availability of raw materials
TECTEG MFR	Calcium/manganese oxide	750 K	12.3 W	Available	800 °C	Environmentally-friendly, low cost, availability of raw materials
TECTEG MFR cascade modules	Calcium/manganese oxides with Bi_2Te_3	435 K	11 W	Available	600 °C	
TECTEG MFR hybrid modules	BiTe-PbTe	320 K	21.7 W	Available	360 °C	Scarce (rare earth), toxicity
Hotblock onboard	Silicon-based alloy	500 K	3.6 W 6 g	Available	600 °C	Environmentally-friendly, low cost, availability of raw materials
Romny scientific	Magnesium silicide			Coming soon	600 °C	Low $/watt target: 1$/W
Alphabet energy	p-type tetrahedrites n-type magnesium silicide	300 K	9.2 W	Available	600 °C	Tetrahedrite is a naturally-occurring p-type mineral
OTEGO CDT	Organic TEG	Small temperature gradients		Coming soon	130 °C	Environmentally-friendly, low cost, easily scalable

Reproduced from Champier [45]

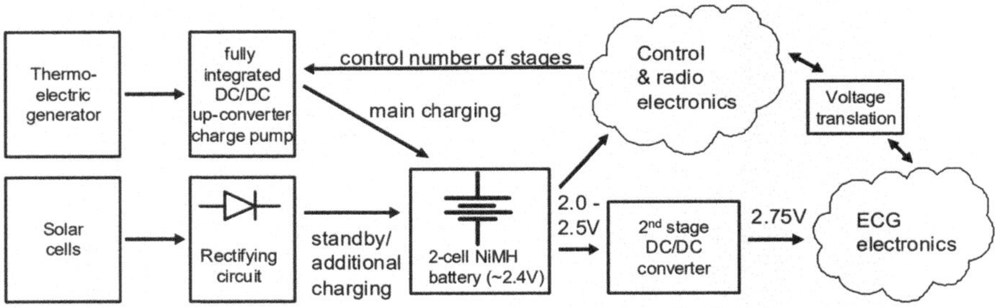

Fig. 17 TEG-PV energy harvesting setup to power ECE electronics. (Reproduced from [47])

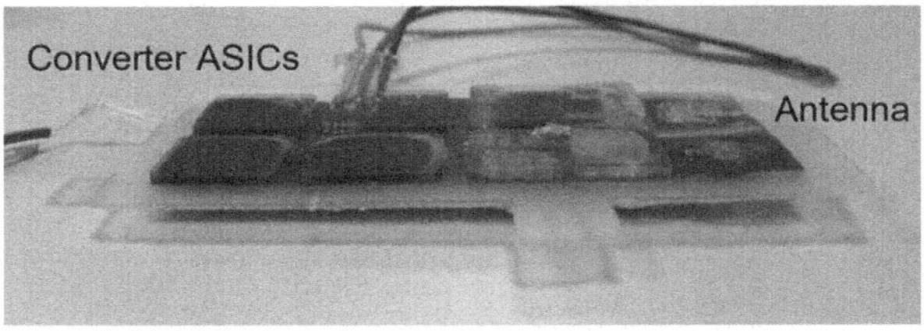

Fig. 18 Flexible electronics integrated with cloths. (Reproduced from [47])

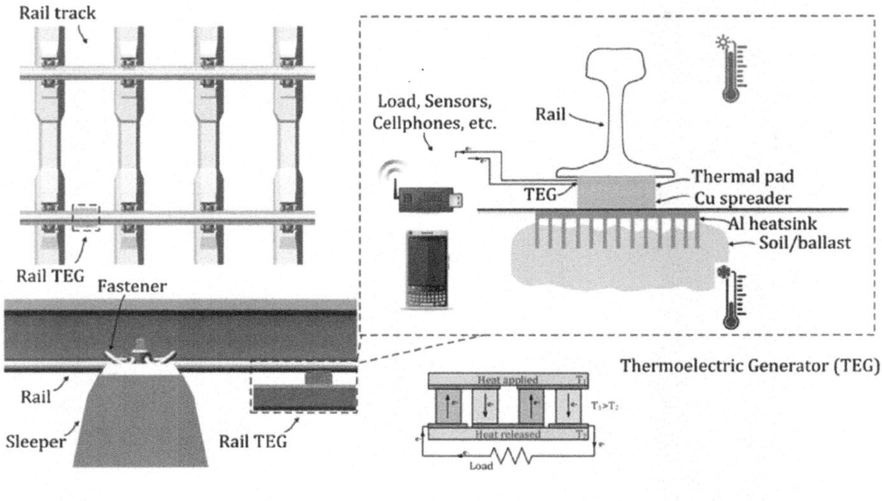

Fig. 19 Illustration of thermoelectric energy harvesting by temperature gradient of railway infrastructure. (Reproduced from [49])

increased to 31.68 mW at a T_D of 29.2 °C. The efficiency of harvesting and conversion was 60%. The harvested energy was used to charge a 900 maH battery and took about 8 h to charge it to 50% (Fig. 19).

A novel "thermoelectric energy harvesting system" (TEHS) (Fig. 20) was developed to harvest thermal energy to supply energy for the onboard electronics of high-

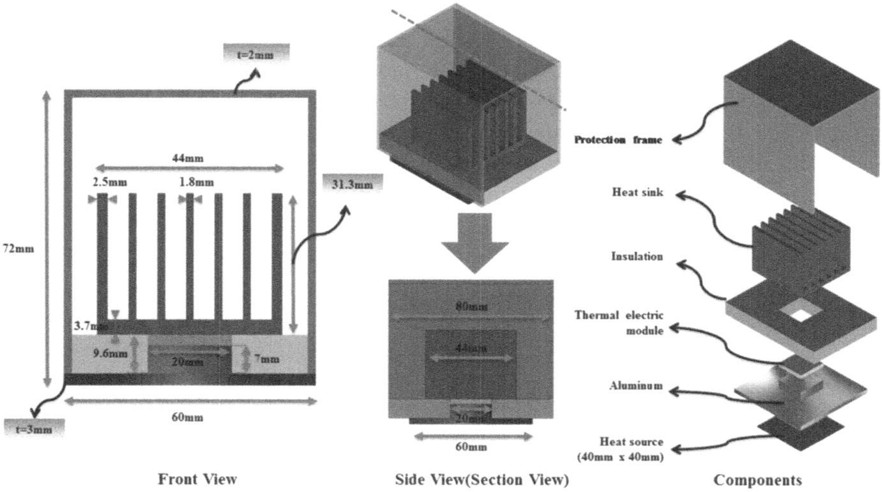

Fig. 20 Illustration of the thermoelectric energy harvesting system (TEHS) for application on axle housing of rolling stock. (Reproduced from [50])

Fig. 21 (**a**) HEMU-430X, Korean next-generation high-speed train, (**b**) TEHS mounted on an axle bearing housing of HEMU-430X train, (**c**) onboard data measurement setup. (Reproduced from [50])

speed vehicles [50]. The module consists of a complete setup to dissipate the heat transferred from the hot junction to the cold junction. The TEHS was placed on an axle-box bearing (Fig. 21) which will have a temperature difference from the surrounding air, and this difference could be used to harvest the energy. Two types of thermoelectric modules (TEM) (similar to TEG) 128A1030 (Peltron) and TK-1-3-S (HTRD) were used for testing. From the Peltron module, P_{max} was 24.6 mW, while from the HTRD module, it was 29.96 mW when the T_d was 30 °C. This power could supply enough energy for operating 2 tri-axis accelerometers or 30 temperature sensors.

Advanced aircraft material's strain can be measured using strain energy sensors in an autonomous sensor system. This system could be powered by TEGs, vibration generators, or solar cells. A TEG-based system was able to generate about 20 mW of power at T_d of 70 °C. The sensor nodes could operate autonomously and communicate using their related protocols [51]. A power management system (PMS) comprises a DC-DC converter with efficiencies of up to 80%. A low-power charge pump is used to lower the start-up voltage to 250 mV from about 800 mV so that the TEG-generated voltage is sufficient to drive the PMS. As the voltage of the storage unit varies with the load, a load-matching circuit was employed.

3 Summary

The purpose of various energy harvesters was to power small electronic systems or to store energy for those electronic systems for later use. For RF energy harvesting, a single-diode rectifier offered better efficiency. Rectenna-antenna offered higher efficiencies at lower RF powers. Power harvested by Archimedean spiral antenna with half-wave Cockcroft-Walton circuit was maximum in LTE and GSM bands of up to 8%. However, overall system efficiency was only 30%. Half-wave Cockcroft-Walton circuit paired with a slotted patch antenna could achieve 68% efficiency for 2.45 GHz frequency. Acoustic energy harvesters with various cavities are used to harvest energy from sound, in combination with organic and inorganic piezoelectric materials to produce powers up to 60 mW/m^2. Lithium niobate produced less electricity than PZT but it has more conversion efficiency. Perovskite structured materials can be tailored for a multitude of applications. The loading and unloading cycles can be used to generate electricity using piezoelectric material. Vibration energy can be harvested using cantilever-based structures combined with electromagnetic generators as well as piezoelectric devices. The power that can be harvested from vibrational energy can be up to 4.9 mW. The fluid flow through the pipes can be used to generate powers up to 12,566 mW using mini turbines, and the oscillations resulted in power generation of about 0.2 μW. Light energy harvesting was done using PV cell and the efficiencies ranged up to 89%. The maximum power harnessed in the indoor environment was about 72.74 μW. Thermal energy from various sources resulted in maximum harnessed power of 31.68 mW at a T_D of 29.2 °C. To harvest thermal energy, proper heat conduction and dissipation mechanism needs to be provided similar to the TEHS. All the energy harvesting needs RF-DC or DC-DC boost converters for utilizing the harvester energy. A supercapacitor is one alternative to the regular Li-based rechargeable battery, with better recharging cycles. MPPT protocol could be implemented to improve the efficiency of the conversion.

References

1. Kim KK, Choi J, Ko SH (2021) Energy harvesting untethered soft electronic devices. Adv Healthc Mater 10(17):2002286
2. Bosso N, Magelli M, Zampieri N (2021) Application of low-power energy harvesting solutions in the railway field: a review. Veh Syst Dyn 59(6):841–871
3. Hesham R, Soltan A, Madian A (2021) Energy harvesting schemes for wearable devices. AEU Int J Electron Commun 138:153888
4. Pozo B, Garate JI, Araujo JA, Ferreiro S (2019) Energy harvesting technologies and equivalent electronic structural models. Electronics 8(5):486
5. Li Z, Xin C, Peng Y, Wang M, Luo J, Xie S, Pu H (2021) Power density improvement of piezoelectric energy harvesters via a novel hybridization scheme with electromagnetic transduction. Micromachines (Basel) 12(7):803
6. Leonov V (2011) Human heat generator for energy scavenging with wearable thermopiles. Sens Transducers 126(3):1
7. Shen S, Chiu C-Y, Murch RD (2017) A dual-port triple-band L-probe microstrip patch rectenna for ambient RF energy harvesting. IEEE Antennas Wirel Propag Lett 16:3071–3074
8. Mrnka M, Vasina P, Kufa M, Hebelka V, Raida Z (2016) The RF energy harvesting antennas operating in commercially deployed frequency bands: a comparative study. Int J Antennas Propag 2016:1–11
9. Sharma T, Saini G (2016) Microstrip antenna array for RF energy harvesting system. IJAIST 5(1). https://doi.org/10.15693/ijaist/2016.v5i1.145-149
10. Ahmed S, Zakaria Z, Husain MN, Yik SW (2015) An array antenna design for RF energy harvesting system. IJAER 10(16):37284–37289
11. Shen S, Chiuand CY, Murch RD (2016) A dual-port triple-band L-probe microstrip patch rectenna for ambient RF energy harvesting. IEEE Antennas Wirel Propag Lett 16:3071–3074
12. Koohestani M, Tissier J, Latrach M (2020) A miniaturized printed rectenna for wireless RF energy harvesting around 2.45 GHz. AEU Int J Electron Commun 127:153478
13. Alex-Amor A, Palomares-Caballero A, Fernández-González JM, Padilla P, Marcos D, Sierra-Castañer M, Esteban J (2019) RF energy harvesting system based on an archimedean spiral antenna for low-power sensor applications. Sensors 19(6):1318
14. Hagerty JA, Helmbrecht FB, McCalpin WH, Zane R, Popovic ZB (2004) Recycling ambient microwave energy with broad-band rectenna arrays. IEEE Trans Microw Theory Tech 52:1014–1024
15. Björkqvist O, Dahlberg O, Silver G, Kolitsidas CI, Quevedo-Teruel O, Jonsson BLG (2018) Wireless sensor network utilizing radio-frequency energy harvesting for smart building applications. IEEE Antennas Propag Mag 2018:60
16. Mattsson M, Kolitsidas CI, Jonsson BLG (2018) Dual-band dual polarized full-wave rectenna based on differential field sampling. IEEE Antennas Wirel Propag Lett 17:956–959
17. Awais Q, Jin Y, Chattha HT, Jamil M, He Q, Khawaja BA (2018) A compact rectenna system with high conversion efficiency for wireless energy harvesting. IEEE Access 6:35857–35866
18. Choi J, Jung I, Kang C-Y (2019) A brief review of sound energy harvesting. Nano Energy 56:169–183
19. Yang J, Chen J, Liu Y, Yang W, Su Y, Wang ZL (2014) Triboelectrification-based organic film nanogenerator for acoustic energy harvesting and self-powered active acoustic sensing. ACS Nano 8(3):2649–2657
20. Izhar, Khan FU (2016) An improved design of helmholtz resonator for acoustic energy harvesting devices. In: International conference on intelligent systems engineering, pp 2–7
21. Iizumi S, Shu K, Tomioka S, Tsujimoto K, Uchida Y, Tomii K, Matsuda T, Nishioka Y (2011) Lead zirconate titanate acoustic energy harvesters utilizing different polarizations on diaphragm. Proc Eng 25:187–190
22. Wang DA, Chang KH (2010) Electromagnetic energy harvesting from flow induced vibration. Microelectron J 41(6):356–364

23. Kim S, Clark WW, Wang QM (2005) Piezoelectric energy harvesting with a clamped circular plate: analysis. J Intell Mater Syst Struct 16:847–854
24. Arnau A (2004) Piezoelectric transducers and applications. Springer, New York
25. Kim S, Clark WW, Wang QM (2005) Piezoelectric energy, harvesting with a clamped circular plate: experimental study. J Intell Mater Syst Struct 16:855–863
26. Platt SR, Farritor S, Haider H (2005) On low-frequency electric power generation with PZT ceramics. IEEE-ASME Trans Mechatron 10:240–252
27. Funasaka T, Furuhata M, Hashimoto Y and Nakamura K (1998) Piezoelectric generator using a LiNbO3 plate with an inverted domain. Ultrasonics Symp. (Sendai), pp 959–62
28. Yoon Y-J, Park W-T, Li KHH, Ng YQ, Song Y (2013) A study of piezoelectric harvesters for low-level vibrations in wireless sensor networks. Int J Precis Eng Manuf 14(7):1257–1262
29. Yang J, Lee M, Park M-J, Jung S-Y, Kim J (2015) A 2.5-V, 160–μJ-output piezoelectric energy harvester and power management IC for batteryless wireless switch (BWS) applications. In Proceedings of Symposium VLSI Circuits (VLSI Circuits), pp C282–C283
30. Lee J, Choi B (2014) Development of a piezoelectric energy harvesting system for implementing wireless sensors on the tires. Energy Convers Manag 78:32–38
31. Wang W, Vinco A, Pavlov N, Wang N, Hayes M, O'Mathuna C (2013) A rotating machine acoustic emission monitoring system powered by multi-source energy harvester. In:Proceedings of the 1st International Workshop on Energy Neutral Sensing Systems (ENSSys), New York, pp 5:1–5:6
32. Sari I, Balkan T, Külah H (2010) An electromagnetic micro power generator for low-frequency environmental vibrations based on the frequency upconversion technique. J Microelectromech Syst 19(1):14–27
33. Gao MY, Wang P, Cao Y, Chen R, Liu C (2016) A rail-borne piezoelectric transducer for energy harvesting of railway vibration. J Vibroengineering 18(7):4647–4663
34. Wang DA, Ko HH (2010) Piezoelectric energy harvesting from flow-induced vibration. J Micromech Microeng 20(2):025019
35. Azevedo JAR, Santos FES (2012) Energy harvesting from wind and water for autonomous wireless sensor nodes. IET Circ Devic Syst 6(6):413–420
36. Nasiri A, Zabalawi SA, Mandic G (2009) Indoor power harvesting using photovoltaic cells for low-power applications. IEEE Trans Ind Electron 56(11):4502–4509
37. Firas Saaduldeen Ahmed et al. (2020) Power quality improvement by using multiple sources of PV and battery for DSTATCOM based on coordinated design. In: IOP conference series: materials science and engineering, vol 745, pp 012025
38. Damodaran R, Rincón-Mora GA (2013, May 19–23) Battery-assisted and photovoltaic-sourced switched-inductor CMOS harvesting charger–supply. In: IEEE's International Symposium on Circuits and Systems (ISCAS), Beijing
39. Yu H, Yue Q (2016) Indoor light energy harvesting system for energy-aware wireless sensor node. Energy Procedia 16:1027–1032
40. Jabbar H, Jeong T (2022) Ambient light energy harvesting and numerical modeling of non-linear phenomena. Appl Sci 12(4):2068
41. Iyer V, Bayati E, Nandakumar R, Majumdar A, Gollakota S (2018) Charging a smartphone across a room using lasers. Proc ACM Interact Mob Wearable Ubiquitous Technol 1(4):1–21
42. Gajanur NR, Singh A, Jain A (2016, October). Solar powered railway f monitoring system. IEEE International Conference on Power and Renewable Energy (ICPRE), pp 190–194
43. Toh WY, Tan YK, Koh WS, Siek L (2014) Autonomous wearable sensor nodes with flexible energy harvesting. IEEE Sensors J 14(7):2299–2306
44. Silva-Leon J, Cioncolini A, Nabawy MR, Revell A, Kennaugh A (2019) Simultaneous wind and solar energy harvesting with inverted flags. Appl Energy 1(239):846–858
45. Champier D (2017) Thermoelectric generators: a review of applications. Energy Convers Manag 140:167–181
46. Ken Brazier (n.d.). https://en.wikipedia.org/wiki/File:Thermoelectric_Generator_Diagram.svg

47. Leonov V, Torfs T, Van Hoof C, Vullers RJM (2009) Smart wireless sensors integrated in clothing: an electrocardiography system in a shirt powered using human body heat. Sens Transducers 107(8):154–176
48. Leonov V (2013) Thermoelectric energy harvesting of human body heat for wearable sensors. IEEE Sensors J 13(6):2284–2291. https://doi.org/10.1109/JSEN.2013.2252526
49. Gao M, Su C, Cong J et al (2019) Harvesting thermoelectric energy from railway track. Energy 180:315–329
50. Ahn D, Choi K (2018) Performance evaluation of thermoelectric energy harvesting system on operating rolling stock. Micromachines 9(7):359
51. Becker T, Kluge M, Schalk J et al (2009) Autonomous sensor nodes for aircraft structural health monitoring. IEEE Sensors J 9(11):1589–1595

Piezo-energy Harvesting and Application Prospects

Shivam Tiwari and Pralay Maiti

1 Introduction

Energy harvesting is one of the most promising and growing technologies these days due its sustainability and environmental viability. It is a technique to harness the waste or unused energy to useful and productive energy. There are different types of works which generate some or other kind of waste or unattended source of energies, which on methodically harnessing and converting provides considerable amount of output energy which can be utilized for small or low-power-based works. The methodology to convert the waste or unattended energy to fruitful electrical output is termed as energy harvesting. Energy harvesting has received both academic and industrial attention due to its self-powered and autonomous mechanism to produce energy which does not generate any harmful byproduct. The ability to produce independent or autonomous self-powered energy has provided a solution to design the devices which are no longer reliant on the batteries. The battery-less device will be a boon for the world as it does not require any replacement or recharging and reduce the disposal of the waste batteries which is not environmental friendly [10, 65, 80]. The constant depletion of the natural resources, results in rise in the fuel prices and generating a concerning perspective for the coming future. Conventional energy sources have the major contribution toward the energy supply to the world, but the limited resources and harmful effects it causes due to its use in the form of pollution and global warming have pushed to alternative sources which are green and sustainable. Several approaches have been developed as an alternative and sustainable source of energy like solar energy [15], fuel cell [64, 65], and

S. Tiwari (✉) · P. Maiti
School of Materials Science and Technology, Indian Institute of Technology (Banaras Hindu University) Varanasi, Varanasi, UP, India
e-mail: shivamtiwari.rs.mst17@itbhu.ac.in; pmaiti.mst@itbhu.ac.in

A. Nella et al. (eds.), *Energy Harvesting Trends for Low Power Compact Electronic Devices*, EAI/Springer Innovations in Communication and Computing,
https://doi.org/10.1007/978-3-031-35965-1_2

Table 1 Average power generated from different human body activities [66, 72]

Human body activities	Average power
Breathing	0.83 W
Blood pressure	0.93 W
Finger motion	6.9–19 mW
Exhalation	1.0 W
Body heat	2.4–4.8 W
Arm motion	60 W
Footfall	67 W

supercapacitor [41]. Energy harvesting is one such approach which has the potential to minimize the constraint in the alternate energy sources due to its unharming byproduct and better scalability.

Energy harvesting is a process that uses a variety of techniques and resources to produce sustainable energy. Several alternate renewable sources are present like solar energy, wind energy, geothermal, and hydrothermal which are highly abundant in the nature and a great potential for the sustainable energy harvesting. Even though they possess higher efficiency, yet certain limitations like availability at every location, distribution, and economically viable make it confined in its use. Other common sources of the energy are radio frequency (RF), acoustic waves, and vibrational energy. The most prevalent energy source is one that relies on mechanical or vibrational energy. Any object that is vibrating or moving has the potential to be an energy source. Movements brought on by the flow of air, water, or turbines may have the potential to be an energy source [53]. It is possible to generate energy from the movement of cars, the vibrations of bridges and tunnels, the operation of industrial machinery like compressors and motors, or even the human body. Human body is a rich source of energy which gets unused or ignored. Movements of hands and limbs, heartbeat, and mouth movement while speaking or chewing are some of the waste mechanical sources which generate substantial amount of power that can be utilized for miniatured device-based applications [60]. Starner [72] presented an overview of the energy generated from the different human body movements.

Table 1 presents the estimated energy generation from different human body activities [66, 72]. Of the different energy sources, mechanical or vibrational energy sources are of prime importance due to its scalability, durability, and high power density. Three techniques are typically used to convert mechanical or vibrational energy into meaningful electrical output: electrostatic [57], electromagnetic [26], and piezoelectric [67] or triboelectric [50]. Between the parallel plate capacitors, where the electrical energy is being stored, electrostatic energy is harvested. By anchoring one of the plates and allowing the other plates to move due to external motion, which modifies the capacitors' characteristics, the energy is gathered (separation between plates or the area). The main drawback of these is that the energy harvester must move with the assistance of an external source [3]. By combining magnetism with electric current to create a magnetic field, electromagnetic energy can be harvested. Magnetic energy is produced when an electric current is delivered because it causes magnetization. This energy harvesting method's main flaw is that

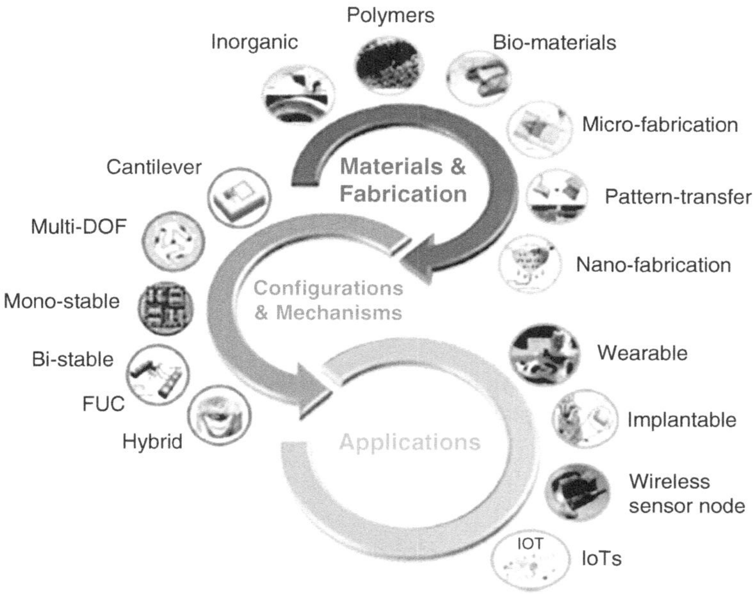

Fig. 1 A broad/extensive outline of the basic materials, configurations, and applications of piezoelectric energy harvesting [51]

its output energy is rather low, necessitating amplification in order to store energy [8, 26]. A more recent development in energy harvesting is the triboelectric energy harvesting which utilizes the concept of generating electrical charges on frictional contact. These processes produce energy with a high conversion efficiency and density. The long-term endurance of these approaches, however, is one of their main drawbacks [14]. The notion of piezoelectricity, in which stress is applied to produce electrical energy, is used in the process of piezoelectric energy harvesting (PEH). These mechanisms provide energy at a higher density and can operate in a variety of ways, including longitudinal, transverse, and piezotronic modes. The absence of moving components, simple construction, portability, scalability, quick production method, affordability, and environmental safety are the primary benefits of piezoelectric energy harvesting [24, 39] (Fig. 1).

2 Piezoelectric Energy Harvesting (PEH)

2.1 Piezoelectric Effect

Piezoelectricity is a term which relates to the generation of electrical energy when a material is being deformed or subjected to stress. The term piezoelectricity is derived from the Greek word "piezen" that implies the application of pressure or being squeezed [27]. The piezoelectric phenomenon was first introduced in 1880 by Jacques and Pierre Curie. The piezoelectric phenomenon is a technique wherein applying mechanical stress causes the production of electrical energy or applying electrical energy causes strain in the materials. In general, piezoelectric effect is of two types:

Direct piezoelectric effect: When an external load or stress is subjected over the material, an output is generated as electrical energy:

$$D = d\sigma + \in E \tag{1}$$

Indirect piezoelectric effect: An output strain is developed when an input electrical energy is provided or applied to the material:

$$\varepsilon = s\sigma + dE \tag{2}$$

The symbols D and E denote the electric displacement and electric field, respectively, whereas the symbols σ, ε, and $\in$ imply the stress, strain, and permittivity, respectively, while the notations s and d represent elastic compliance and the piezoelectric coefficients.

The organization of the ions in the unit cell and the crystalline solids are the important features that can be used to describe the fundamentals of piezoelectricity. Equivalent positive and negative charges are present in the crystalline structure, where they stay neutralized along the polar axis. When an external mechanical stress or load is applied, the periodic arrangement within the crystal is disturbed, which causes the dipoles to be arranged improperly, resulting to the production of a dipole moment and, as a result, the generation of an electric charge. The crystals' symmetry plays a key role in shaping the system's piezoelectric characteristic. There are around 32 classes of crystallographic units, 21 of which are noncentrosymmetric and 20 of which have characteristics associated with piezoelectricity. The absence of symmetry in the group of ions in the materials causes the growth of permanent dipoles, which are related to the piezo-response [11, 80].

2.2 Mechanism

The piezoelectric energy harvesting follows the mechanism of the direct piezoelectric effect. The creation of the net dipole moment in solids and the piezoelectric effect are related phenomena. The alignment of the molecular dipoles, a change in the configuration of the dipoles, or changes in the environment of the dipoles can all cause a change in polarization when external stress is applied to the material. Free charges, such as ions or electrons, are drawn to the charged surfaces of the material because it has a polarization with aligned dipoles and domains, which results in the presence of a charge at each of the material's surfaces. When an external stress is applied over the materials, the structure gets deformed which results in the change in the arrangement of charges which disrupts the symmetry in the material. Due to the lack of symmetry in the ion arrangement, these crystalline solids have a net dipole moment, which produces the piezoelectric effect. When no stress is applied over the material, no charge generation occurs; hence no flow of electron takes place [6, 79].

The charge produced (Q) across the opposite faces of the material is defined as [10].

$$Q = \mathrm{d}_{33} \times \Delta\sigma \times A \tag{3}$$

where the term A denotes the area of the piezoelectric material, d_{33} is the piezoelectric coefficient, and $\Delta\sigma$ is the stress applied over the material.

Under open circuit conditions having infinite load impedance,

$$Q = CV \tag{4}$$

where the capacitance (C) can be given as

$$C = A \times \varepsilon^{\mathrm{T}}{}_{33}/h \tag{5}$$

The voltage generated can be written as [80]

$$V = \left(\mathrm{d}_{33}/\varepsilon^{\mathrm{T}}{}_{33}\right) \times h \times \Delta\sigma \tag{6}$$

The term h is the thickness, while $\varepsilon^{\mathrm{T}}{}_{33}$ denotes the permittivity at constant stress in the direction of polarization. The term ($\mathrm{d}_{33}/\varepsilon^{\mathrm{T}}{}_{33}$) is referred to as the figure of merit (FoM) which evaluates the performance of a material in case of vibration-based energy harvesting off resonance. With the load impedance which is zero under short-circuit conditions, the associated current (I) can be given as [80].

$$I = \mathrm{d}_{33} \times A \times \frac{\Delta\sigma}{\Delta t} \tag{7}$$

2.3 Mechanical Energy Harvester

Energy harvesting from the mechanical sources converts the energy generated from the movement or vibration of the source to productive electrical energy. Generally the energy generated from the mechanical-based energy harvesting is of small-scale range which is used to operate or utilize in low-power electronics which is a viable technique to replace the batteries. Basically there are mainly two ways to harness energy from mechanical sources: inertial and kinematic. The ability to harvest inertial energy depends on a mass' resistance to acceleration. When the base is moved or displaced, a resultant force is produced in a mass-spring system. These systems have a single point of connection to the base and are commonly utilized for vibration harvesting. A vibration is created in the mass-spring system when the base moves, and this vibration can be used to produce electrical energy. Since the vibration amplitude of a system at resonance might be much greater than the amplitude of the base movement, the vibration amplitude is not just proportional to the base amplitude. Kinematic energy harvesting associates the energy harvester directly to the relative motion of several energy sources. Inertia and resonance are not used in kinematic energy harvesting methods. They are connected at multiple points because the strain in the harvester is directly attached to a branch of the source [10].

Mechanical energy harvesters are the fundamental materials which transform the input mechanical energy to electrical energy. The energy harvester is based on different mechanism as per the functional applications. The basic operation of an energy harvester can be demonstrated based on the following points [10]:

- Basic features of the energy source.
- The way of transformation of energy from source to the energy harvester.
- Electromechanical conversion.
- Process of transformation of the energy source to the electrical load from the energy harvester.

Piezoelectric energy harvester employs the force of an external force operating on piezoelectric components to produce energy. This technology is typically employed to transform ambient waste energy into useable electrical energy. The direct piezoelectric effect serves as the foundation for the piezoelectric energy harvester's mechanism. Charges will be created on the surface of the materials proportionally to the forces placed on the harvester. The charges cause current to flow through the load when connected to an external circuit. In this process, the piezoelectric material is essentially a source of voltage, current, charge, or power. The terms "energy scavenger" and "power generator" are occasionally used to describe piezoelectric energy harvesters [48]. Although the magnitude of the applied stress or strain also counts, the efficiency of energy harvesting is closely tied to the piezoelectric coefficients. The connection between the mechanical source and the piezoelectric material has a significant impact on how well energy is harvested because of this. The energy output is also influenced by the piezoelectric material's ability to endure an applied force or repeatedly undergo a recoverable strain. It is essential for kinematic energy harvesters because these limitations on the materials' strength and elasticity may have a greater impact on the efficiency of energy harvesting than just the piezoelectric coefficients [10]. The direction of the applied stress in relation to the polar axis will have an impact on the energy harvesting performance because the majority of piezoelectric materials for energy harvesting have a well-defined polar axis. The applied stress onto the piezoelectric material can be in two

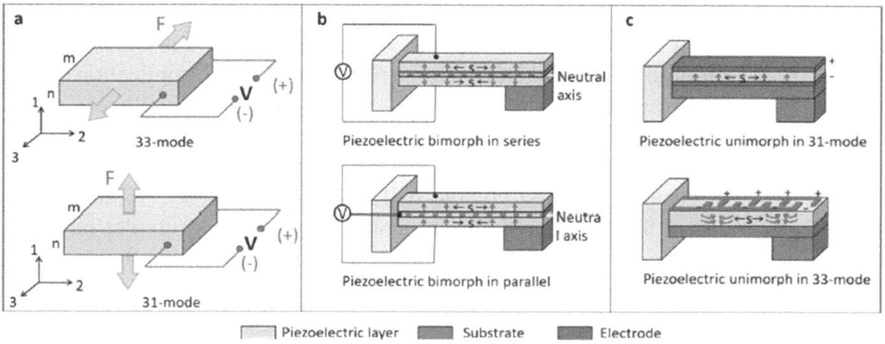

Fig. 2 (**a**) Piezoelectric material representation in the 31 and 33 mode; (**b**) series and parallel connection of 31-mode bimorph cantilever; and (**c**) unimorph cantilever configuration of 31 and 33 mode [51]

directions: three-direction (polar axis) and one-direction (right angle to polar axis). Based on the two directions of the applied stress, two important configurations for piezoelectric energy harvesting occur, namely, "31" and "33." There can be other possible configurations but these two hold most of the situations. In the 33 mode, compressive stress and strain are applied perpendicular to the polar axis and parallel to the three-direction, respectively, while voltage is generated along the same axis, whereas in the 31 mode, the direction of the produced voltage is at a right angle to the direction of the applied force [51]. To extract the energy for a piezoelectric energy harvester, different device configurations are being considered. The cantilever is one of the most widely used PEH designs because it can induce a substantial amount of mechanical strain during vibration within the piezoelectric material, which is especially beneficial for mechanical energy harvesting from vibrations. More notably, in comparison to the other designs, the fundamental flexural modes of a cantilever have a resonance frequency that is somewhat lower [51]. The vibration source is secured to the cantilever's root end, while the opposite end is attached to a mass. While the base accelerates, the cantilever is bent by the tip mass' inertia. Simply producing equal and opposite strains on the inside and outside of the bend is enough to bend a piezoelectric element. As a result, no net current is generated. To act as a generator, the piezoelectric layer needs to be moved away from the neutral axis. This is frequently accomplished by joining two piezoelectric layers with opposing poles or by fastening the piezoelectric substance to an elastic layer made of a different material. The unimorph or the bimorph are terms used to describe these arrangements [10] (Fig. 2).

The most popular type of bimorph cantilever is made up of two distinct piezoelectric sheets that have been joined together using a center shim or flange. With the top layer of the parts in tension and the bottom layer in compression, or vice versa, the structure is intended to function in the bending mode. The piezoelectric phenomenon is the basis for how this structure generates electric charge. For each layer to produce an accumulated current or voltage, the top and bottom layers are poled either in the same direction or in the opposite direction. This is known as parallel or series poling [81]. A cantilever with only one active layer and one dormant layer is known as a unimorph. If the active layer is piezoelectric, the application of an electric field may cause deformation in that layer. The cantilever bends as a result of this distortion [25]. Unambiguously, the type of piezoelectric mode has a significant impact on the benefits and performance of a piezoelectric unimorph cantilever. It is anticipated that the generated voltage will be higher for the 33-mode energy harvester than that of the 31-mode device by considering that both modes have the same design characteristics. Additionally, for the 31-mode device and the 33-mode device, respectively, the voltage generation is proportional to the spacing between the electrode fingers and the top and bottom electrodes. Micro-electromechanical system (MEMS) PZT cantilevers based on the 31 mode and the 33 mode have been compared by Kim et al. [40]. They came to the conclusion that the 33-mode device can produce higher voltage and power as compared to the 31 mode.

3 Piezoelectric Materials

Piezoelectric materials are smart materials that produce a solid-state conversion of electrical and mechanical energy. The electric polarization of the piezoelectric material changes when external stress or load is applied, which induces electric charges into the system. The ability to generate potential or charge is therefore necessary for piezoelectric materials in order to produce an electrical signal under the effect of an external load. Performance of energy harvesting is significantly influenced by how well piezoelectric materials couple to mechanical energy sources. One of the most important characteristics for energy harvesting applications is the piezoelectric material's capacity to withstand the applied mechanical stress and produce an efficient output [10, 80].

Some important parameters are being introduced to define the efficacy of the piezoelectric materials. The common parameters which are being considered are piezoelectric charge coefficient (d_{ij}), piezoelectric voltage constant (g_{ij}), dielectric permittivity (ε_r), and electromechanical coupling factor (k^2). The generated electric polarization in a material per unit of external mechanical load or stress applied is known as the piezoelectric charge constant, also known as the strain constant, or it can also be described as the strain produced on the application of the external field. Normally, it is expressed as coulombs per newton. The direction of the produced polarization or the direction of the applied field strength is indicated by the first subscript in the notation d_{ij}. The second notation, on the other hand, represents the direction of the subjected stress or the direction of the strain caused in the material. Piezoelectric voltage constant is the generated electric field divided by the mechanical stress or strain applied or the mechanical strain experienced divided by the applied electric displacement. Typically, it is expressed as volts per newton. The direction of the generated electric field or the direction of the electric displacement is represented by the first subscript in the notation *gij*, whereas the direction of the

Table 2 Common piezoelectric materials and their properties [10, 56, 76, 80]

Materials	Piezoelectric coefficient (pC/N)	Relative permittivity	Electromechanical coupling factor
$LiNbO_3$	6	28.7	0.23
AlN	5	12	0.23
ZnO	12.4	11	0.48
$BaTiO_3$	149	1200	0.49
PZT	593	3400	0.75
PVDF	−24 to −34	6–12	0.2
P(VDF-co-TrFE)	−25 to 40	18	0.29
P(VDF-co-HFP)	−24	11	0.36
Polyamide 11	4	5	–
Polyhydroxybutyrate	1.6–2	2–3.5	–

applied stress or the direction of the induced strain in the material is represented by the second subscript. Electromechanical coupling factor is the ratio of mechanical energy that has been stored to applied electrical field or vice versa. Being the ratio of energy, these units have no dimensions. Dielectric permittivity states the ability of the piezoelectric materials to hold the electric charges. It is formulated as the ratio of the permittivity of the material to free space [6, 7, 63].

The first piezoelectric material was natural quartz which was discovered by Jacques and Pierre Curie in 1880. Post the discovery of quartz, several piezoelectric materials have been introduced to the world [51, 55]. Basically there are four important classes of the piezoelectric material such as single crystals, ceramics, polymers, and composites. The choice of piezoelectric material depends on the specific property for particular application along with the functionality of that application in which it is to be used [56]. Table 2 provides a list of common piezoelectric materials with its properties. The common categories of piezoelectric materials are discussed below.

3.1 Piezoelectric Single Crystals

The piezoelectric single crystals consist of positive and negative ions are arranged in a periodic pattern over the structure, which produces a dipole moment as a result of the correct dipole alignment. These characteristics of the single crystal result in highly effective piezoelectric qualities, which are applied to sensors and actuators. Lithium tantalite, lithium niobate, and quartz are three common single-crystal piezomaterials. These materials are effective for piezoelectric-based applications because they have significant polarization. In addition to these benefits, piezoelectric single crystals nevertheless have certain drawbacks, such as difficult fabrication processes and expensive synthesis, which restrict its use in a variety of applications. Because there are no ceramic grain boundaries, single crystals tend to be fragile by nature. Additionally, in an environment with a greater electric field, a single crystal tends to lose its piezoelectric capabilities [34, 49]. Chen et al. [12] reviewed the lead-based single crystals like PMN-PT and PZN-PT which possess very high d_{33} and k_{33} values which makes them a suitable material for energy harvester and applicable in the medical imaging field, sonar and actuators.

3.2 Piezoelectric Ceramics

The microscopic crystals that make up piezoelectric ceramics have random crystal orientations, and they don't become piezoelectric until they undergo a process called polarization, which usually involves applying a strong electrical field to align the crystal orientations. These are the inorganic polycrystalline materials, which are composed of different single crystals with unique chemical compositions. These materials are rigid, stiff, and brittle with very high dielectric constants and piezoelectric coefficients. However, the substance is deemed inappropriate in several specific application domains due to its toxicity. Ceramics made of non-ferroelectric materials are another potential source of piezoelectricity. These materials are valuable for biological sensors and semiconducting applications despite having lower piezoelectric values than their ferroelectric siblings because of their high selectivity and sensitivity. Common piezoceramics are lead zirconate titanate (PZT), lead magnesium niobate-lead titanate (PMN-PT), barium titanate ($BaTiO_3$), zinc oxide (ZnO), and others [4, 17, 58, 69, 74]. Chen et al. [13] prepared PZT-based nanofibers by electrospinning techniques and utilized it for mechanical energy harvesting. The fabricated device was able to generate output voltage and power of 1.6 V and 0.03 μW, respectively. Koka et al. [43] prepared the vertically aligned ferroelectric $BaTiO_3$ nanowire (NW) arrays and traditional ZnO NW arrays in a unique way to create nano-electromechanical system (NEMS) vibrational energy harvesters with resonance frequencies under 200 Hz. The peak power density generated was around 6.27 mW/cm^3 for the $BaTiO_3$ NW arrays which was around 16-fold higher as compared to the ZnO NW.

3.3 Piezoelectric Polymers

Piezopolymers are the class of materials which possess long range order, superior flexibility, and mechanical durability. In specific applications, electroactive polymers clearly outperform ceramics due to their simple processing method at low temperatures, low density, better flexibility, and mechanical robustness, including toughness and high strain to failure. For implantable harvesters and sensors, there are also advantages in terms of biocompatibility. Due to their high levels of polarization, ferroelectric polymers with net molecular dipole moments are of interest for energy storage and harvesting. By applying an electric field, changing the temperature, mechanically stretching the material, and interacting with nanoparticles, the molecular dipoles in these materials can spontaneously polarize and orient. Polyvinylidene fluoride or PVDF and its copolymers are the most efficient and largely used piezoelectric polymer in the field of energy harvesting applications. Other piezopolymers are polyamide, cellulose and its derivatives, chitin, and polyacrylonitrile [59, 69, 80]. Hanninen et al. [29] prepared a free standing biodegradable film from chitosan and cellulose and analyzed its piezoelectric sensitivity. The average sensitivities for the prepared films were (21.03 ± 6.80) pC/N, (11.23 ± 6.76) pC/N, (8.74 ± 3.66) pC/N, and (5.99 ± 2.91) pC/N for pure chitosan, chitosan/cellulose nanocrystals (CNC), chitosan/neutralized cellulose nanocrystals, and chitosan/cellulose nano-fibril (CNF) films, respectively. Polyamides are other piezoelectric polymers which show good level of piezoelectric properties. Takase et al. [75] prepared the films of Nylon-11 and Nylon-7 from hot pressing and observed the piezoelectric strain constant (d_{31}) and electromechanical coupling coefficient (k_{31}). The maximum values for d_{31} for Nylon-11 and Nylon-7 were around 14 pC/N and 17 pC/N, respectively, while the k_{31} values were 0.054 and 0.049 for Nylon-7 and Nylon-11, respectively.

PVDF is one of the noteworthy polymers in the field of PEH applications. The piezoelectricity in PVDF was first reported by Kawai in 1969 [38]. It is a semicrystalline polymer with the alternating $-CH_2$ and $-CF_2$ groups in its chain. The difference in the electronegativity between the elements leads to generation of net dipole moment in the polymer which results in the piezoelectric property. PVDF mainly possess three distinct phases: nonpolar α-phase, semipolar γ-phase, and polar β-phase. PVDF consists of predominant nonpolar α-phase which does not show piezoelectric property. The efficacy of PVDF largely depends on the transformation of the non-piezo-phase to electroactive phase. Piezoelectric active β-phase is one of the important phases for the energy harvesting applications due to its high spontaneous polarization [32]. The conversion of the nonpolar phase of PVDF to piezoactive phase can be obtained from poling [31], stretching [68], electrospinning [33], or incorporation of electroactive fillers to the PVDF matrix [76–78]. Gaur et al. [19] prepared PVDF film and studied the effect of mechanical stretching and poling for energy harvesting applications. PVDF film showed better electroactive phase formation on stretching and poling as compared to pristine film. Szewczyk et al. [73] prepared PVDF fibers from electrospinning process and demonstrated the influence of ambient relative humidity and voltage polarity in the electrospun fibers for energy harvesting applications. A flexible nanogenerator based on direct piezoelectric effect has been created employing a thin spin-coated PVDF-TrFE film as the functional layer. The nanogenerator is capable of producing an open-circuit voltage of up to 7 V and a short-circuit current of 58 nA at a current density of 0.56 $\mu A/cm^2$ [62].

Apart from these classified materials, there exist some bio-based piezoelectric materials which have gained wide attention due to its biocompatibility and simple approach for better energy harvesting. However the accountability and its long time durability is a major drawback due to the lifetime of these bio-piezoelectric materials. Some common bio-based piezoelectric materials are silk, DNA, and bone [51]. Maiti et al. [54] prepared a flexible piezoelectric nanogenerator from bio-waste onion skin which generated an output voltage and current of around 18 V and 166 nA, respectively. Karan et al. [36] developed an egg-shell membrane-based bio-piezoelectric nanogenerator which showed piezoelectric coefficient (d_{33}) of 23.7 pC/N and maximum output voltage of around 26.4 V.

3.4 *Composites*

A composite is made up of two or more components with similar or different traits that combine to enhance the system's physical and chemical properties. The major goal of this is to produce a substance that combines the best attributes of its constituent elements to provide modified and sustainable properties that are ideal for the applications being sought after. Better mechanical, electrical, and thermal qualities are offered by composite materials, along with reduced density and resistance to corrosion. The composite is used in a variety of industries, including packaging, infrastructure, medicine, aerospace, and many more. In polymer composites, the flexibility and improved toughness of the polymer matrix are used to complement the benefits of abrasion resistant, chemical resistant, higher impact characteristics, and corrosion resistant [5, 20, 70]. Addition of electroactive fillers to the matrix leading to higher piezoelectric values is an advanced process for efficient piezoelectric energy harvesting. Ceramics being one of the prominent piezoelectric materials with higher piezoelectric properties is restricted to be used in certain applications due to its brittleness and toxic nature. But incorporation of the ceramic materials to the piezo-active polymers inhibits the drawbacks and hence provides better flexibility and durability which can have wider applicability. Wankhade et al. [84] prepared PVDF-PZT composite at different ratio of PZT and observed its piezoelectric properties. Maximum output voltage and power density of 55 V and 36 μWcm^{-2}, respectively, is achieved for PVDF-PZT (30 wt. %) which is much higher as compared to the neat PVDF film.

According to Hua et al. [30], solid-state mechanochemically made polyamide-11 (PA11)/$BaTiO_3$ piezoelectric composites exhibited decreasing order with rising frequency and rising order of dielectric constant as the volume fraction of $BaTiO_3$ rose. Nanoparticle inclusion also changed the crystallization behavior, improving thermal stability. The produced composite's piezoelectric coefficients similarly showed a linear connection with filler volume percentage, with d_{33} increasing from 1.1 to 6.6 pC/N when the $BaTiO_3$ loading was increased from 5% to 50%.

Carbon-based fillers have gained wider attention due to their better thermal, mechanical, and piezoelectric properties. Bhavanasi et al. [9] prepared a bilayer film using PVDF-TrFE and GO to assess the composite's piezoelectric behavior. There was a definite demonstration of how poling affected the performance of EH. In comparison to the non-poled bilayer film, which produced a voltage of 0.3 V against a mechanical stress of 0.32 MPa, the bilayer film with poled PVDF-TrFE and GO generated a maximum output voltage of 4.3 V. Therefore, the film's poling results in greater dipole alignment, which raises the electromechanical responsiveness. The

Table 3 Piezoelectric output of different piezoelectric materials

Materials	Processing	Piezoelectric output	References
PVDF-nanoclay nanofibers (5 wt.%)	Electrospinning	Free vibration damping test = 2.76 V Finger tapping = 0.83 V	[86]
PVDF/graphene nanoplatelets (0.1, 1, 3 5 wt.%)	Electrospinning	V = 7.9 V, I = 4.5 μA (at 0.2 MPa)	[1]
PVDF-TrFE+ MgO nanoparticles (2–8%)	Solution casting	Vpp = 4.54 V; d_{33} = ~ 65 pm/V	[71]
PVDF-cellulose nanocrystal nanofibers	Electrospinning	Maximum voltage output of ~60 V@ 2 wt % CNC	[18]
Spider silk		V = 21.3 V; I = ~0.68 μA	[37]
M13 bacteriophage		V = 400 mV, I = 6 nA; d_{33} = 7.8 pm/V	[46]
PDMS+ BZT-BCT Nanorods	Poling	V = 0.8 V, I = 7 nA (finger pressing)	[44]
PDMS+MWCNT +$ZnSnO_3$		V = 40 V, I = ~0.4 μA, P = ~10.8 μWcm^{-3}	[2]
FaPbBr3+ PDMS (5, 20, 35, 50%)	Spin coated	V = 8.5 V; d_{33} = 25 pm/V	[16]
PVDF-AlO-rGO		V = 36 V, I = ~0.8 μA, P = 27.97μWcm^{-3}	[35]

poled bilayer film's maximum power output was approximately 4.41 $\mu W/cm^2$, which was superior to the PVDF-TrFE film made of virgin polymer, which had a power density of 1.77 $\mu W/cm^2$ at a load resistance of 1 MΩ. PVDF-CNT fibers were created by Wu et al. using an electrospinning technique. When CNT nanoparticles were added, the polar phase significantly increased from 79% to 89% of the aligned PVDF fibers. The pure PVDF random fibers had a piezoelectric coefficient (d_{33}) of 16.8 ± 1.4 pC/N, which rose to 31.3 ± 2.1 pC/N for the aligned PVDF-CNT fibers. The electroactive fillers cause nonpolar phases to change into polar or semipolar phases [85]. Several works have been demonstrated where the bio-waste materials have been incorporated with the polymers to produce enhanced piezoelectric properties. Kumar et al. [45] prepared a composite of PVDF and fish scale which showed improved piezoelectric properties in comparison to the pristine PVDF. Gaur et al. [21] prepared PVDF-orange peel composite and demonstrated the piezoelectric energy harvesting applications. In other work, pomegranate peel-based PVDF hybrid was prepared which showed increase in the polar nature and generated better electromechanical response [23].

Different approaches have been implemented to generate piezoelectric energy for energy harvesting applications. Table 3 presents the piezoelectric output of different materials for energy harvesting applications. Park et al. [61] prepared a metal/PVDF/metal layer-based device which was then used to study the energy harvesting effect of the tensile stress over the prepared material. The device generated sustainable energy output against acoustic vibrations, and also the device was able to store charge in terms of voltage against external vibrations. In other work, tri-functional device based on all powder-processing methods by using piezoelectric material as dielectric layer and ZnS powder as phosphor layer was prepared. With the help of an external electric power source, the constructed gadget produced piezoelectric sheet sound and electroluminescent (EL) light from phosphor. It also captured reverse-piezoelectric energy that could be used to generate EL light. By just using mechanical pressure and no external electric power to the device, the piezoelectric-induced electroluminescence effect was established as an energy harvesting application [47]. Han et al. [28] developed a controlled, nonlinear buckling process to convert lithographically defined two-dimensional patterns of electrodes and thin films of piezoelectric polymers into sophisticated 3D piezoelectric microsystems. The prepared 3D piezoelectric polymer material was successfully demonstrated for vibration energy harvesting, biomedical implants, and robotic interfaces applications.

4 Applications

Piezoelectric materials have attracted widespread attention in the field of energy harvesting application due to their better durability, scalability, and cost-effectiveness. The fabrication of the energy harvesting device to harness the waste or unused energies to productive electrical output has been implemented in several fields of applications like transducers, actuators, sensors, healthcare, robotics, and others [22, 76]. Human body is a great source of energy which when harvested can be of great technological importance. Liu et al. [52] reviewed piezoelectric-based energy harvesting from the movement or activities of the different upper limb parts of human body. The role of different piezoelectric materials like polymers, ceramics, and composites was analyzed, and its importance in piezoelectric energy harvesting through upper limb parts was discussed. An improved breathability and sensitivity PVDF-based force sensor was created by Wang et al. Utilizing the electrospinning method, the PVDF fabric was prepared. ITO-coated glass and plastic film were used as the electrodes for designing the device, and polymer fabric was inserted in the space between them. In comparison to the other films, the fabric manufactured at 12 kV potential, 12 wt. % concentrations, and flow rate of 0.02 ml/min had produced the maximum β-phase content and shown the highest sensitivity of 42 mV/N. Additionally, the device was able to produce a strong signal at several frequencies [82]. In order to transform the mechanical energy produced by the human body into useful electrical energy, a piezoelectric nanogenerator was constructed using 2D boron nitride nanosheets (BNNs) as piezoelectric active component and PDMS as flexible part. Maximum output voltage of 22 V, output current of 75 nA, and 40 μW of maximum power were generated. The constructed gadget was applied to numerous human body parts, including the neck, elbow, wrist, knee, and foot, and a significant amount of energy was generated [42]. Additionally, Zhou et al. demonstrated a 3D-printed piezoelectric energy harvester for analyzing gait. Devices are made of materials from kirigami-shaped PVDF-TrFE polymers improved with $BaTiO_3$ nanoparticles. The gadget can be attached and stretched up to 300% into

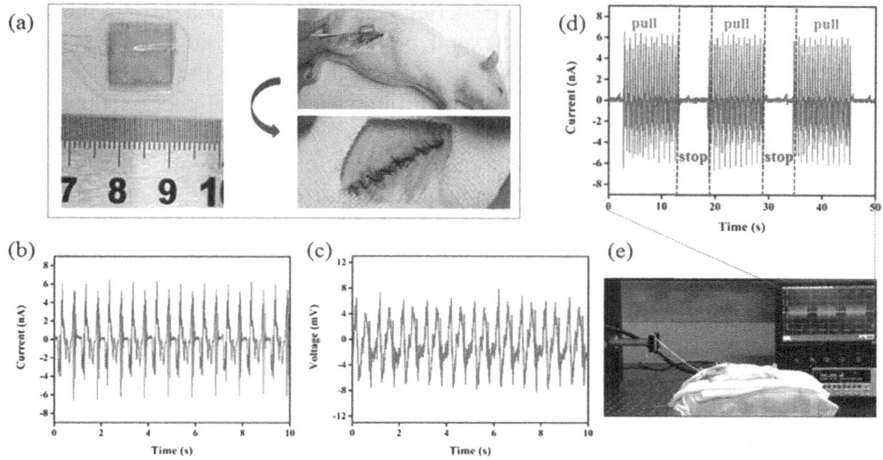

Fig. 3 (**a**) Pictorial view of the prepared P(VDF-TrFE) nanofiber scaffold (before in vivo study) and illustration of the implantation of the material in the SD rat (upper part) and the subsequent image after the suturing process (lower part); (**b**) current output; (**c**) voltage output; (**d**) current output from the pulling-releasing; and (**e**) view of the experimental setup [83]

textiles that may be worn, like the harvester and sock gait sensor. The output voltage of 6 V and current density of 2 μA/cm^2 are generated from the prepared material. The maximum power density of 1.4 μW/cm^2 is produced at density at 10^7 Ω [87]. Wang et al. [83] prepared the electrospun fibers from PVDF-TrFE through electrospinning process, and the fiber morphology was optimized by varying the solution concentration and the collector distance. The maximum output voltage, output current, and piezoelectric coefficient (d_{31}) achieved for the poled scaffold are around 1.5 V, 52.5 nA, and 15.73 pC/N, respectively. From the application perspective, the material was implanted under the legs of Sprague Dawley (SD) rats, and the movement of the rats was observed in terms of generation of output voltage and current. The maximum output voltage and current generated from the pulling of legs with a linear motor through wire to demonstrate the daily activity of the rat were around 6 mV and 6 nA, respectively. Hence the prepared scaffold demonstrated a good applicability of the piezoelectric material in field of biomedical and healthcare applications (Fig. 3).

5 Summary and Future Perspective

Energy harvesting from the waste mechanical energy has provided a new alternative energy source which is sustainable and environmentally viable. PEH using the smart piezoelectric materials has provided a new perspective to the field with better scalability and energy efficiency. Piezoelectric materials have garnered wide consideration from the world owing to its rich benefits and applicability in the field of energy harvesting applications. Several piezoelectric materials like polymers, ceramics, and composites have provided wide applicability to the field of energy harvesting. Harnessing of energy from the different mechanical energy sources to productive electrical energy has been implemented using different materials and structures which have provided an alternate solution to the conventional materials. This chapter provides an insight to the mechanism of piezoelectricity and the common piezoelectric materials which have been used widely for the energy harvesting applications. The chapter briefly describes the common device configurations and the piezoelectric materials being used for the piezoelectric-based energy harvesting applications. The piezoelectric-based energy harvesting scavenges energy mainly from the strain or vibration which varies according to the source. Harnessing each type of waste or unused energy is a challenge which is being considered, and required steps are being implemented for enhancing the productivity of the output energy in terms of increasing the device size or focusing on area where maximum energy is generated and has potential to be utilized for energy generation for sustainable future. Effective power management using modified circuits and development of piezoelectric materials with high piezoelectric coefficients would provide wide applicability of these materials.

References

1. Abolhasani MM, Shirvanimoghaddam K, Naebe M (2017) PVDF/graphene composite nanofibers with enhanced piezoelectric performance for development of robust nanogenerators. Compos Sci Technol 138:49–56. https://doi.org/10.1016/j.compscitech.2016.11.017
2. Alam MM et al (2015) Lead-free ZnSnO3/MWCNTs-based self-poled flexible hybrid nanogenerator for piezoelectric power generation. Nanotechnology:165403. https://doi.org/10.1088/0957-4484/26/16/165403
3. Aljadiri RT (2017) Electrostatic energy harvesting systems: a better understanding of their sustainability electrostatic energy harvesting systems: a better understanding of their sustainability a null space approach for complete and over-complete blind source separation of autoregressive source signals view project blind source separation of ECG and speech signals using Null space idempotent (NSITM) algorithm view project. https://doi.org/10.18178/JOCET.2017.5.5.407
4. Anton SR, Sodano HA (2007) A review of power harvesting using piezoelectric materials (2003–2006), 1. https://doi.org/10.1088/0964-1726/16/3/R01
5. Avila AF et al (2003) A dual analysis for recycled particulate composites: linking micro- and macro-mechanics. Mater Charact 50(4–5):281–291. https://doi.org/10.1016/S1044-5803(03)00124-4
6. Batra AK et al (2016) Piezoelectric power harvesting devices: an overview. Adv Sci Eng Med 8(1):1–12. https://doi.org/10.1166/ASEM.2016.1819
7. Baur C et al (2014) Advances in piezoelectric polymer composites for vibrational energy harvesting. In: ACS symposium series, vol. 1161. https://doi.org/10.1021/BK-2014-1161.CH001/ASSET/IMAGES/MEDIUM/BK-2013-00468N_G069.GIF
8. Beeby SP, O'Donnell T (2009) Electromagnetic energy harvesting. In: Energy harvesting technologies. Springer, Boston, pp 129–161. https://doi.org/10.1007/978-0-387-76464-1_5

9. Bhavanasi V et al (2016) Enhanced piezoelectric energy harvesting performance of flexible PVDF-TrFE Bilayer films with graphene oxide. ACS Appl Mater Interfaces 8(1):521–529. https://doi.org/10.1021/acsami.5b09502
10. Bowen CR et al (2014) Piezoelectric and ferroelectric materials and structures for energy harvesting applications. Energy Environ Sci 7(1):25–44. https://doi.org/10.1039/c3ee42454e
11. Caliò R et al (2014) Piezoelectric energy harvesting solutions. Sensors 14(3):4755–4790. https://doi.org/10.3390/S140304755
12. Chen J, Panda R (2005) Commercialization of piezoelectric single crystals for medical imaging applications. In: IEEE Ultrasonics Symposium, 2005 (Vol. 1, pp. 235–240). IEEE.
13. Chen X et al (2010) 1.6 v nanogenerator for mechanical energy harvesting using PZT nanofibers. Nano Lett 10(6):2133–2137. https://doi.org/10.1021/nl100812k
14. Covaci C, Gontean A (2020) Piezoelectric energy harvesting solutions: a review. Sensors 20(12):3512. https://doi.org/10.3390/S20123512
15. Devabhaktuni V et al (2013) Solar energy: trends and enabling technologies. Renew Sust Energ Rev 19:555–564. https://doi.org/10.1016/j.rser.2012.11.024
16. Ding R et al (2016) Flexible piezoelectric nanocomposite generators based on formamidinium lead halide perovskite nanoparticles. Adv Funct Mater 26(42):7708–7716. https://doi.org/10.1002/ADFM.201602634
17. Fang J et al (2013) Enhanced mechanical energy harvesting using needleless electrospun poly (vinylidene fluoride) nanofibre webs. Energy Environ Sci 6(7):2196–2202. https://doi.org/10.1039/C3EE24230G
18. Fu R et al (2017) Improved piezoelectric properties of electrospun poly(vinylidene fluoride) fibers blended with cellulose nanocrystals. Mater Lett 187:86–88. https://doi.org/10.1016/J.MATLET.2016.10.068
19. Gaur A et al (2018) Efficient energy harvesting using processed poly(vinylidene fluoride) nanogenerator. ACS Appl Energy Mater 1(7):3019–3024. https://doi.org/10.1021/acsaem.8b00483
20. Gaur A et al (2019) Polymer composites for structural, device, and biomedical applications. In: Materials science and technology. https://doi.org/10.1002/9783527603978.mst0455
21. Gaur A et al (2020a) Bio-waste orange peel and polymer hybrid for efficient energy harvesting. Energy Rep 6:490–496. https://doi.org/10.1016/j.egyr.2020.02.020
22. Gaur A et al (2020b) Flexible, lead-free nanogenerators using poly(vinylidene fluoride) nanocomposites. Energy Fuel 34(5):6239–6244. https://doi.org/10.1021/acs.energyfuels.0c01143
23. Gaur A et al (2020c) Polymer biowaste hybrid for enhanced piezoelectric energy harvesting. ACS Appl Electron Mater 2(5):1426–1432. https://doi.org/10.1021/ACSAELM.0C00197/SUPPL_FILE/EL0C00197_SI_002.AVI
24. Ghazanfarian J, Mohammadi MM, Uchino K (2021) Piezoelectric energy harvesting: a systematic review of reviews. Actuators 10(12):312. https://doi.org/10.3390/ACT10120312
25. Giri AM, Ali SF, Arockiarajan A (2022) Piezoelectric unimorph and bimorph cantilever configurations: design guidelines and strain assessment. Smart Mater Struct 31(3):035003. https://doi.org/10.1088/1361-665X/AC47D5
26. Glynne-Jones P et al (2004) An electromagnetic, vibration-powered generator for intelligent sensor systems. Sensors Actuators A Phys 110(1–3):344–349. https://doi.org/10.1016/J.SNA.2003.09.045
27. Haertling GH (1999) Ferroelectric ceramics: history and technology. J Am Ceram Soc 82(4): 797–818. https://doi.org/10.1111/J.1151-2916.1999.TB01840.X
28. Han M et al (2019) Three-dimensional piezoelectric polymer microsystems for vibrational energy harvesting, robotic interfaces and biomedical implants. Nat Electron 2(1):26–35. https://doi.org/10.1038/s41928-018-0189-7
29. Hänninen A et al (2016) Piezoelectric sensitivity of a layered film of chitosan and cellulose nanocrystals. Procedia Eng 168:1176–1179. https://doi.org/10.1016/J.PROENG.2016.11.397

30. Hua Z, Shi X, Chen Y (2019) Preparation, structure, and property of highly filled polyamide 11/BaTiO3 piezoelectric composites prepared through solid-state mechanochemical method. Polym Compos 40(S1):E177–E185. https://doi.org/10.1002/PC.24566
31. Jin Z et al (2021) Influences of poling temperature and elongation ratio on PVDF-HFP piezoelectric films. Nanotechnol Rev 10(1):1009–1017. https://doi.org/10.1515/NTREV-2021-0070/DOWNLOADASSET/SUPPL/NTREV-2021-0070_SM.PDF
32. Kabir E et al (2017) Pure β-phase formation in polyvinylidene fluoride (PVDF)-carbon nanotube composites. J Phys D Appl Phys 50(16):163002. https://doi.org/10.1088/1361-6463/AA5F85
33. Kalimuldina G et al (2020) A review of piezoelectric PVDF film by electrospinning and its applications. Sensors 20(18):5214. https://doi.org/10.3390/S20185214
34. Karami MA et al (2011) Experimental and analytical parametric study of single-crystal unimorph beams for vibration energy harvesting. IEEE Trans Ultrason Ferroelectr Freq Control 58(7):1508–1520. https://doi.org/10.1109/TUFFC.2011.1969
35. Karan SK et al (2016) An approach to design highly durable piezoelectric nanogenerator based on self-poled PVDF/AlO-rGO flexible nanocomposite with high power density and energy conversion efficiency. Adv Energy Mater 6(20):1–12. https://doi.org/10.1002/aenm.201601016
36. Karan SK, Maiti S, Paria S et al (2018a) A new insight towards eggshell membrane as high energy conversion efficient bio-piezoelectric energy harvester. Mater Today Energy 9:114–125. https://doi.org/10.1016/j.mtener.2018.05.006
37. Karan SK, Maiti S, Kwon O et al (2018b) Nature driven spider silk as high energy conversion efficient bio-piezoelectric nanogenerator. Nano Energy 49(April):655–666. https://doi.org/10.1016/j.nanoen.2018.05.014
38. Kawai H (1969) The piezoelectricity of poly (vinylidene fluoride). Jpn J Appl Phys 8(7): 975–976. https://doi.org/10.1143/JJAP.8.975/META
39. Khalid S et al (2019) A review of human-powered energy harvesting for smart electronics: recent progress and challenges. Int J Precis Eng Manuf Green Technol 6(4):821–851. https://doi.org/10.1007/S40684-019-00144-Y
40. Kim SB et al (2013) Comparison of MEMS PZT cantilevers based on d31 and d 33 modes for vibration energy harvesting. J Microelectromech Syst 22(1):26–33. https://doi.org/10.1109/JMEMS.2012.2213069
41. Kim BK et al (2015) Electrochemical supercapacitors for energy storage and conversion. In: Handbook of clean energy systems. Wiley, pp 1–25. https://doi.org/10.1002/9781118991978.HCES112
42. Kim KB et al (2018) Transparent and flexible piezoelectric sensor for detecting human movement with a boron nitride nanosheet (BNNS). Nano Energy 54:91–98. https://doi.org/10.1016/J.NANOEN.2018.09.056
43. Koka A, Zhou Z, Sodano HA (2013) Vertically aligned BaTiO3 nanowire arrays for energy harvesting. Energy Environ Sci 7(1):288–296. https://doi.org/10.1039/C3EE42540A
44. Kou Y et al (2017) Bio-compatible BCTZ-based piezoelectric nanogenerator as energy harvester and waterdrop counter. Ceram Int 43(9):6666–6670. https://doi.org/10.1016/J.CERAMINT.2017.02.028
45. Kumar C et al (2019) Bio-waste polymer hybrid as induced piezoelectric material with high energy harvesting efficiency. Compos Commun 11:56–61. https://doi.org/10.1016/J.COCO.2018.11.004
46. Lee BY et al (2012) Virus-based piezoelectric energy generation. Nat Nanotechnol 7(6): 351–356. https://doi.org/10.1038/nnano.2012.69
47. Lee S et al (2018) Multifunctional device based on phosphor-piezoelectric PZT: lighting, speaking, and mechanical energy harvesting. Sci Rep 8(1):2–8. https://doi.org/10.1038/s41598-017-18571-9
48. Li T, Lee PS (2022) Piezoelectric energy harvesting technology: from materials, structures, to applications. Small Structures 3(3):2100128. https://doi.org/10.1002/SSTR.202100128

49. Li H, Tian C, Deng ZD (2014) Energy harvesting from low frequency applications using piezoelectric materials. Appl Phys Rev 1(4):041301. https://doi.org/10.1063/1.4900845
50. Lin Z et al (2017) Triboelectric nanogenerator enabled body sensor network for self-powered human heart-rate monitoring. ACS Nano 11(9):8830–8837. https://doi.org/10.1021/ACSNANO.7B02975/SUPPL_FILE/NN7B02975_SI_003.AVI
51. Liu H et al (2018) A comprehensive review on piezoelectric energy harvesting technology: materials, mechanisms, and applications. Appl Phys Rev 5(4). https://doi.org/10.1063/1.5074184
52. Liu Y et al (2021) Piezoelectric energy harvesting for self-powered wearable upper limb applications. Nano Select 2(8):1459–1479. https://doi.org/10.1002/NANO.202000242
53. Mahapatra SD et al (2021) Piezoelectric materials for energy harvesting and sensing applications: roadmap for future smart materials. Adv Sci 8(17):2100864. https://doi.org/10.1002/ADVS.202100864
54. Maiti S et al (2017) Bio-waste onion skin as an innovative nature-driven piezoelectric material with high energy conversion efficiency. Nano Energy 42(October):282–293. https://doi.org/10.1016/j.nanoen.2017.10.041
55. Manbachi A, Cobbold RSC (2011) Development and application of piezoelectric materials for ultrasound generation and detection. Ultrasound 19(4):187–196. https://doi.org/10.1258/ULT.2011.011027
56. Mishra S et al (2019) Advances in piezoelectric polymer composites for energy harvesting applications: a systematic review. Macromol Mater Eng 304(1):1–25. https://doi.org/10.1002/mame.201800463
57. Mitcheson PD et al (2004) MEMS electrostatic micropower generator for low frequency operation. Sensors Actuators A Phys 115(2–3):523–529. https://doi.org/10.1016/J.SNA.2004.04.026
58. Muralt P (2000) Ferroelectric thin films for micro-sensors and actuators: a review. J Micromech Microeng 10(2):136. https://doi.org/10.1088/0960-1317/10/2/307
59. Nalwa H (1995) Ferroelectric polymers: chemistry: physics, and applications. Available at: https://books.google.co.in/books?hl=en&lr=&id=_M8B6ISHxFUC&oi=fnd&pg=PP17&dq=Nalwa,+H.+S.+(Ed.).+(1995).+Ferroelectric+polymers:+chemistry:+physics,+and+applications.+CRC+Press&ots=Q_Uu6VdZN2&sig=kpZD7EmI60GkCZborkGAwahDdRs. Accessed 18 Feb 2021
60. Paradiso JA, Starner T (2005) Energy scavenging for mobile and wireless electronics. IEEE Pervasive Comput 4(1):18–27. https://doi.org/10.1109/MPRV.2005.9
61. Park S et al (2017) Energy harvesting efficiency of piezoelectric polymer film with graphene and metal electrodes. Sci Rep 7(1):1–8. https://doi.org/10.1038/s41598-017-17791-3
62. Pi Z et al (2014) Flexible piezoelectric nanogenerator made of poly(vinylidenefluoride-co-trifluoroethylene) (PVDF-TrFE) thin film. Nano Energy 7:33–41. https://doi.org/10.1016/j.nanoen.2014.04.016
63. Power Enhancement in Piezoelectric Energy Harvesting (n.d.). Available at: https://www.diva-portal.org/smash/record.jsf?pid=diva2%3A941980&dswid=-8579. Accessed 25 Jul 2022
64. Prakash O et al (2022a) Dehydrohalogenated poly(vinylidene fluoride)-based anion exchange membranes for fuel cell applications. Mater Today Chem 23:100640. https://doi.org/10.1016/J.MTCHEM.2021.100640
65. Prakash O, Tiwari S, Maiti P (2022b) Fluoropolymers and their nanohybrids as energy materials: application to fuel cells and energy harvesting. ACS Omega. https://doi.org/10.1021/ACSOMEGA.2C04774
66. Qi Y, McAlpine MC (2010) Nanotechnology-enabled flexible and biocompatible energy harvesting. Energy Environ Sci 3(9):1275–1285. https://doi.org/10.1039/C0EE00137F
67. Qian F et al (2018) Investigation of piezoelectric energy harvesting from human walking. J Phys Conf Ser 1052(1):012113. https://doi.org/10.1088/1742-6596/1052/1/012113

68. Sencadas V, Gregorio R, Lanceros-Méndez S (2009) α to β phase transformation and microstructural changes of PVDF films induced by uniaxial stretch. J Macromol Sci Part B Phys 48(3):514–525. https://doi.org/10.1080/00222340902837527
69. Setter N et al (2006) Ferroelectric thin films: review of materials, properties, and applications. J Appl Phys 100(5):051606. https://doi.org/10.1063/1.2336999
70. Shaw A et al (2010) A critical reliability evaluation of fibre reinforced composite materials based on probabilistic micro and macro-mechanical analysis. Compos Part B 41(6):446–453. https://doi.org/10.1016/J.COMPOSITESB.2010.05.005
71. Singh D, Choudhary A, Garg A (2018) Flexible and robust piezoelectric polymer nanocomposites based energy harvesters. ACS Appl Mater Interfaces 10(3):2793–2800. https://doi.org/10.1021/ACSAMI.7B16973/ASSET/IMAGES/LARGE/AM-2017-169737_0009.JPEG
72. Starner T (1996) Human-powered wearable computing. IBM Syst J 35(3–4):618–629. https://doi.org/10.1147/SJ.353.0618
73. Szewczyk PK et al (2020) Enhanced piezoelectricity of electrospun polyvinylidene fluoride fibers for energy harvesting. ACS Appl Mater Interfaces 12(11):13575–13583. https://doi.org/10.1021/acsami.0c02578
74. Tadigadapa S, Mateti K (2009) Piezoelectric MEMS sensors: state-of-the-art and perspectives. Meas Sci Technol 20(9):092001. https://doi.org/10.1088/0957-0233/20/9/092001
75. Takase Y et al (1991) High-temperature characteristics of Nylon-11 and Nylon-7 piezoelectrics. Macromolecules 24(25):6644–6652. https://doi.org/10.1021/MA00025A014/ASSET/MA00025A014.FP.PNG_V03
76. Tiwari S, Maiti P (2022) 2D materials–polymer composites for developing piezoelectric energy-harvesting devices. In: Engineered polymer nanocomposites for energy harvesting applications, pp 99–128. https://doi.org/10.1016/B978-0-12-824155-4.00001-6
77. Tiwari S et al (2019) Enhanced piezoelectric response in nanoclay induced electrospun PVDF nanofibers for energy harvesting. Energy 171:485–492. https://doi.org/10.1016/j.energy.2019.01.043
78. Tiwari S et al (2021) Ionic liquid-based electrospun polymer nanohybrid for energy harvesting. ACS Appl Electron Mater 3(6):2738–2747. https://doi.org/10.1021/ACSAELM.1C00307
79. Tong C (2019) Emerging materials for energy harvesting. In: Introduction to materials for advanced energy systems, pp 719–817. https://doi.org/10.1007/978-3-319-98002-7_11
80. Wan C, Bowen CR (2017) Multiscale-structuring of polyvinylidene fluoride for energy harvesting: the impact of molecular-, micro- and macro-structure. J Mater Chem A:3091–3128. https://doi.org/10.1039/c6ta09590a
81. Wang QM et al (1999) Theoretical analysis of the sensor effect of cantilever piezoelectric benders. J Appl Phys 85(3):1702. https://doi.org/10.1063/1.369314
82. Wang YR et al (2011) A flexible piezoelectric force sensor based on PVDF fabrics. Smart Mater Struct 20(4):045009. https://doi.org/10.1088/0964-1726/20/4/045009
83. Wang A et al (2018) Piezoelectric nanofibrous scaffolds as in vivo energy harvesters for modifying fibroblast alignment and proliferation in wound healing. Nano Energy 43:63–71. https://doi.org/10.1016/J.NANOEN.2017.11.023
84. Wankhade SH et al (2020) PVDF–PZT nanohybrid based nanogenerator for energy harvesting applications. Energy Rep 6:358–364. https://doi.org/10.1016/j.egyr.2020.02.003
85. Wu CM, Chou MH, Zeng WY (2018) Piezoelectric response of aligned electrospun polyvinylidene fluoride/carbon nanotube nanofibrous membranes. Nano 8(6):1–13. https://doi.org/10.3390/nano8060420
86. Xin Y et al (2016) Full-fiber piezoelectric sensor by straight PVDF/nanoclay nanofibers. Mater Lett:136–139. https://doi.org/10.1016/j.matlet.2015.09.117
87. Zhou X et al (2020) All 3D-printed stretchable piezoelectric nanogenerator with non-protruding kirigami structure. Nano Energy 72:104676. https://doi.org/10.1016/J.NANOEN.2020.104676

Graphitic Carbon Nitride-Based Dye-Sensitized Solar Cells and Perovskite Solar Cells for Energy Harvesting

Bhanu Chandra Marepally, Maneesh Reddy Venumbaka, Selvakumar Duraisamy, Saravanan Sigamani, D. Hima Bindu, and Vigneswaran Dhasarathan

1 Introduction

Due to rapid industrialization and urbanization, the day-to-day energy demand increased substantially, leading to the fast depletion of the conventional fuel sources (coal, natural gas, oil, etc.). Moreover, the carbon emissions released from the combustion of these fuels result in environmental pollution and climatic imbalances. This created the need for immediate action toward identifying the alternate sources, which are abundant, renewable, and eco-friendly to meet the world energy demand.

B. C. Marepally
Chaitanya Bharathi Institute of Technology, Hyderabad, TS, India
e-mail: bhanuchandram_ece@cbit.ac.in

M. R. Venumbaka (✉)
Chaitanya Bharathi Institute of Technology, Hyderabad, TS, India

KPR Institute of Engg. and Technology, Coimbatore, TN, India

S. Duraisamy (✉)
KPR Institute of Engg. and Technology, Coimbatore, TN, India
e-mail: selvakumar.d@kpriet.ac.in

S. Sigamani
Swarnandhra College of Engineering and Technology, Narsapur, AP, India
e-mail: ssnanotech@swarnandhra.ac.in

D. Hima Bindu
KPR Institute of Engg. and Technology, Coimbatore, TN, India

YSR Engg. College of Yogi Vemana University, Proddatur, AP, India

V. Dhasarathan
University of Hradec Králové, Hradec Králové, Czech Republic
e-mail: vigneswaran.dhasarathan@uhk.cz

A. Nella et al. (eds.), *Energy Harvesting Trends for Low Power Compact Electronic Devices*, EAI/Springer Innovations in Communication and Computing,
https://doi.org/10.1007/978-3-031-35965-1_3

Among the various sources, solar energy becomes a promising alternative because of its high capacity and potential to meet current and future energy demands [1]. It is estimated that the efficient usage of solar irradiance for 1 h across the world is enough to generate 1 year's global energy consumption [2].

Various technologies are being researched and developed to generate energy using sunlight. Among them, solar cells are of key research interest due to their high efficiencies in comparison to other solar technologies. These are categorized into three generations, in which silicon cells belonging to the first generation occupy the major market share, followed by second-generation CIGS solar cells. The remaining types of solar cells are in the lab-scale research and development only. Recently, dye-sensitized solar cells (DSSCs) are in the category of the third-generation solar cells based on thin films and gained much traction and attention from the researchers due to their advantages like simple fabrication process, cost-effectiveness [3, 4], environmental friendliness [5], low weight, easy scalability, etc. In spite of many advantages, there are still some challenges to be overcome, for these cells to become commercially viable such as stability, efficiency, robustness, etc. In addition, another class of cells called perovskite cells took to the front stage and gained sudden popularity due to their exceptional achievement of high efficiencies, within a very short span of their discovery and research. The main challenges of these cells are material structural complexity, toxicity in nature, and stability on exposure to atmospheric conditions, which make them step back a little to rethink and come up with more feasible solutions [2] toward commercialization.

Despite the amazing capabilities of both DSSCs and perovskite cells, they are still facing their own challenges, in order to compete with the well-established silicon solar cell industry. To overcome their challenges, researchers are analyzing the possibility to use different engineered materials using nanotechnology in the preparation of the cell components [6, 7]. The advantage of using engineered nanomaterials is that one can tune the properties of materials at the nanoscale. Different dimensional (3D, 2D, 1D, and 0D) nanostructures like nanosheets, nanoflakes, nanorods, nanotubes, nanospheres, and many other complex nanostructures are being explored and developed based on the materials, like metals, metal oxides, semiconductors, composites, alloys, etc. for the utilization of solar cell applications. Out of which, use of graphitic carbon nitrides (g-C_3N_4) in different fields of solar applications like photocatalytic, photoelectronic, and photovoltaic [8] became more popular, due to the exhibition of superior electronic, optical [9] properties, chemical and thermal stabilities [3], and being metal-free and biocompatible in nature. In this chapter, advancements in applications of DSSCs and perovskite solar cells using g-C_3N_4 are of primary interest along with the different synthesis process of g-C_3N_4, and its structural as well as morphological properties were reviewed.

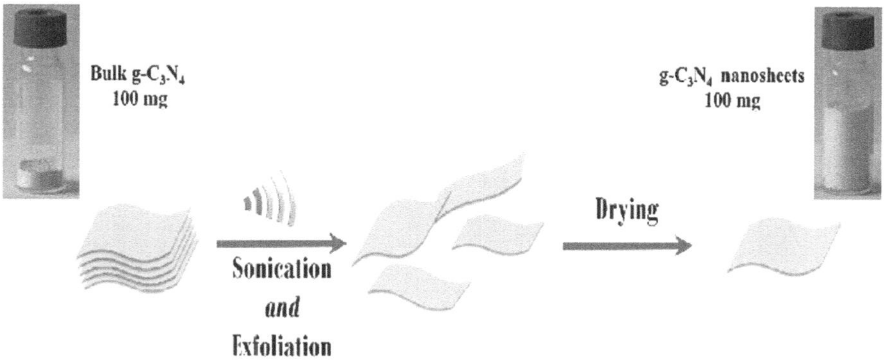

Fig. 1 Ultrasonic-assisted liquid-phase exfoliation method [3]

2 Graphitic Carbon Nitrides: Material Preparation and Structural Properties

In the year 2009, g-C_3N_4 was first reported as a metal-free created photocatalyst for hydrogen (H_2) evolution application by Wang et al. [10]. The structure of g-C_3N_4 is comparable to graphite and has a basic structural unit as triazine (C_3N_3) or heptazine (C_6N_7) [11]. It comprises the elements of carbon C and nitrogen N. It is the more stable form compared to the other types of carbon nitrides [10]. With the bandgap of ~2.7 eV, it has peak absorption of light at ~460 nm [12, 13] which makes g-C_3N_4 as a visible high-energy region light-sensitive material. Different structures of g-C_3N_4, like sheets [3, 14–16], flakes, tubes [9, 17], composite [4, 15, 18, 19], etc., can be synthesized by numerous nitrogen-rich precursors like urea, guanidinium chloride, melamine, thiourea, cyanamide, dicyanamide, guanidine thiocyanate, and ammonium thiocyanate [11, 12, 20] using different routes. A few of them are mentioned below.

Sheetlike structures can be synthesized using techniques like ultrasonic-assisted liquid-phase exfoliation, thermal polycondensation, pyrolysis [14] or thermal polymerization, etc. In the process of ultrasonic-assisted liquid-phase exfoliation, initially, bulk form of g-C_3N_4 is produced by simple heating of melamine [15] or via multistep reactions where melamine is combined with another chemical in presence of a catalyst. Later, the synthesized bulk g-C_3N_4 undergoes sonication for several hours with a constant power and frequency distribution followed by drying. This results in the formation of the sheetlike structures. The steps involved in the formation of nanosheets through the ultrasonic-assisted liquid-phase exfoliation method [3] are shown in Fig. 1.

In the thermal polycondensation process, a small amount of urea is taken and grounded, followed by heating at high temperature (500 °C) through constant rise at 5 °C min^{-1} and ventilated to room ambient [11]. Similarly, in pyrolysis or thermal polymerization, the precursor is heated at higher temperatures with a steady ramp rate for a few hours [14, 22].

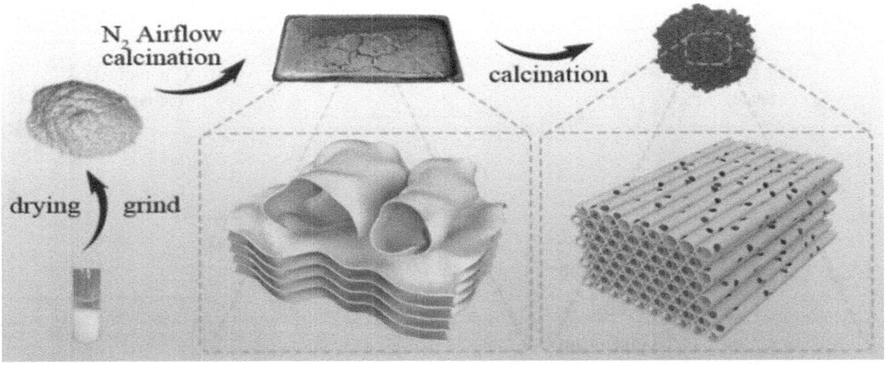

Fig. 2 Formation mechanism of g-C_3N_4 nanotubes [22]

Flake structures also can be synthesized using thermal polycondensation of a precursor followed by ultrasonic exfoliation or decomposition [9, 18]. The obtained flakes were made to shrink by changing their temperature suddenly, which produces thermal stress at their surface, resulting in rolled g-C_3N_4 nanotubes. Another group Xue Li et al. synthesized 1D carbon self-doping of g-C_3N_4 nanotubes through thermal polymerization in saturated or supersaturated urea inside melamine sponge framework for abundant communicating pores [17]. Figure 2 represents the formation mechanism of nanotubes.

g-C_3N_4 films can be formed using dispersive coating method where the bulk g-C_3N_4 is dispersed in a solvent like methanol and acids and then coated onto a substrate followed by drying [23]. Also, by using solvothermal process, in which the substrates were placed in solutions made using precursors (e.g., dissolution of cyanuric chloride and melamine in acetonitrile) and sealed followed by applying higher temperatures around ~450–550 °C. This sets the thermochemical reaction process leading to formation of different nanostructures [23].

Composites can be prepared by either directly mixing the g-C_3N_4 precursors (i.e., urea or melamine) with the desired composite material precursor (TiO_2 nanotubes, nanosheets, nanoparticles) [4, 19] or preparing the bulk g-C_3N_4 initially followed by the addition of composite precursor [16].

Various structural properties of g-C_3N_4 nanostructures can be analyzed using different analytical techniques. To study the information related to the nature of the material and unit cell parameters, we use a technique called X-ray diffraction (XRD). It works based on constructive interference of monochromatic X-rays (e.g., $Cu_{k\alpha} = 0.154$ nm) and a crystalline sample. XRD is not suited for amorphous materials as it may result in broad peaks apart from instrumental broadening [24]. Figure 3 reveals the XRD pattern of g-C_3N_4 nanosheets [23]. The sample had a hexagonal structure belonging to the P63cm space group. The intense XRD peak at 27.5° in the (002) plane represents the graphitic layered stacking through a distance d = 0.326 nm. An additional peak was found at 13.1° in the (100) plane

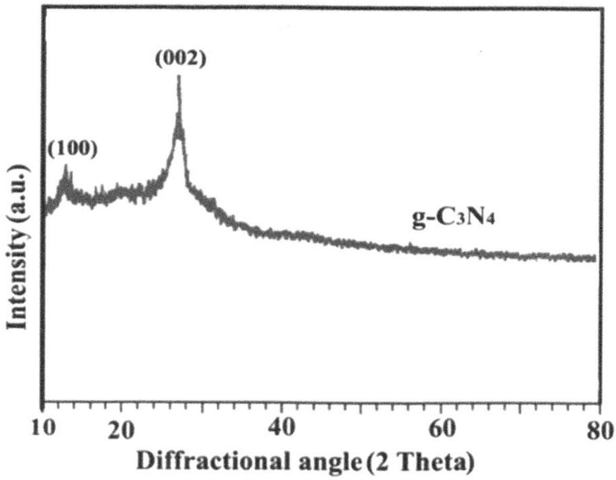

Fig. 3 XRD pattern of the $g\text{-}C_3N_4$ nanosheets [3]

conforming to a distance d = 0.675 nm representing the in-planar repeating units [25].

The crystallite size of the sample was calculated by using Debye-Scherrer equation:

$$D = k\lambda/(\beta \cos \theta)$$

where D is the crystallite size, k is the Scherrer constant (0.9), λ is incident monochromatic radiation wavelength, β is the peak's full width at half maximum, and ⍰ is Bragg's angle. The crystallite size predictable from the (002) plane of the $g\text{-}C_3N_4$ sample is 3.49 nm [26].

Interplanar spacing (d_{hkl}) of the indexed peaks in XRD pattern of the sample was calculated by Bragg's law:

$$d_{hkl} = \lambda/(2\sin \theta)$$

where the subscripts h, k, and l signify the miller indices of the planes, λ is the incident monochromatic radiation wavelength, and ⍰ is Bragg's angle [26].

Morphological analysis was done by scanning electron microscope (SEM) and transmission electron microscope (TEM) techniques. The SEM follows the principle and application of kinetic energy for producing signals with respect to the electrons interaction. TEM applies a beam of electrons in focusing the specimen for image formation. Figure 4a, b reveals the SEM images of $g\text{-}C_3N_4$ nanostructures, which suggested random stacking clubbed morphology with nanorods structure [3]. Figure 4c, d reveals the TEM image of the $g\text{-}C_3N_4$, which suggested a layered and platelet-like morphology [3]. Morphology of the $g\text{-}C_3N_4$ samples may vary like flaky, layered, platelet-like, etc., which depends upon the use of different synthesis

Fig. 4 (**a**, **b**) SEM and (**c**, **d**) TEM image of g-C_3N_4 [3]

methods. The surface area of g-C_3N_4 nanosheets was almost ten times higher than raw g-C_3N_4 [27–29]. High surface area of g-C_3N_4 can be attained by direct polycondensation method without the influence of a template; moreover surface area of the samples either increases or decreases based on the different synthesis methods [30].

3 Graphitic Carbon Nitrides on Dye-Sensitized Solar Application

Dye-sensitized solar cell (DSSC) works on the principle of photosynthesis [31, 32] which involves some key processes, light absorption, charge separation, and charge collection. The main components include anode coated with mesoporous semiconductor loaded with dye, electrolyte for regeneration of dye, and cathode. The structure of DSSC was illustrated in Fig. 5.

The use of g-C_3N_4 in different forms at various components mentioned above was attempted by several groups in order to improve the efficiencies by modifying the properties. Some of them were discussed below.

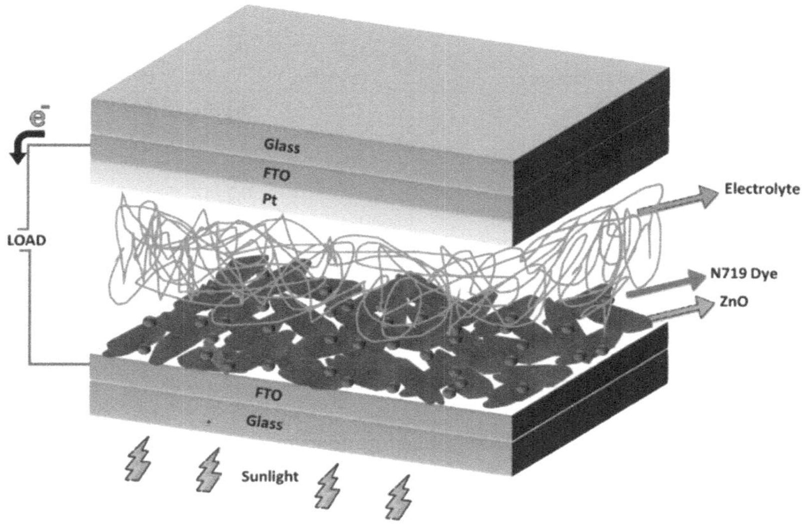

Fig. 5 Schematic of DSSC [33]

3.1 Photoanodes

Huiru Lv et al. compared the efficiencies of DSSCs fabricated using pure TiO_2, which is in anatase form, and TiO_2/g-C_3N_4 composite photoanodes and found 3.87% and 4.51%, respectively. They found that the increase in efficiency is due to reduction of energy barrier of electron transport and also due to broadening in the absorption spectrum of TiO_2 by g-C_3N_4 [4]. Further to investigate the effect of CNTs as photoanodes for the efficiency of DSSCs. Xue Li et al. introduced the synthesized carbon self-doping g-C_3N_4 nanotubes into the TiO_2 to form TiO_2/CNT composite and fabricated DSSC with TiO_2-CNT composite/N719dye/Pt, and its results were related to the DSSC fabricated using pure TiO_2 as a photoanode, and they found the efficiencies are to be 7.4% and 6.3%, respectively. This confirms the introduction of CNTs enhances the efficiency. This is due to improvement in absorption of visible light by g-C_3N_4 and also due to improved charge separation and transfer [17].

R.A. Senthil et al. studied the effect of 2-APY (aminopyridine) in poly(vinylidene fluoride-co-hexafluoropropylene) (PVDF-HFP)/polyethylene oxide (PEO) polymer blend electrolyte and g-C_3N_4/TiO_2 nanocomposite as photoanode on the efficiency of DSSC. From the results, they concluded that incorporation of g-C_3N_4 into the TiO_2 photoanode/pure PVDF-HFP/PEO polymer blend electrolyte constructed DSSC increased the efficiency from 2.46% to 3.17% due to increase in dye absorption and incorporation of 2-APY electrolyte and further enhanced the efficiency to 4.73%, due to the increase in charge transfer among 2-APY and iodide ions [21]. D. Wu et al. studied the effect of g-C_3N_4 coating in the DSSC efficiency by fabricating

Table 1 I–V characteristics of DSSCs based on g-C_3N_4-modified semiconductor photoanodes

References	DSSC photoanode	J_{sc} (mA/cm^2)	V_{oc} (Volts)	FF (%)	Efficiency (%)
Huiru Lv et al.	TiO_2/g-C_3N_4	11.29	0.706	56.58	4.51
Xue Li et al.	TiO_2-CNT composite	17.6	0.00069	0.62	7.4
Senthil et al.	TiO_2/g-C_3N_4	7.43	0.721	0.53	4.73
Wu et al.	ZnO-C_3N_4 (3%)	14.6	0.596	51.7	4.5
J. Xu et al.	TiO_2/g-C_3N_4 (80% wt. ratio of urea)	15.77	0.646	0.720	7.34
Yan et al.	TiO_2/g-C_3N_4	11.76	0.708	0.641	5.34
	g-C_3N_4/ag/TiO_2	12.68	0.715	0.686	6.22
Yuan et al.	TiO_2/g-C_3N_4	16.73	0.82	58.69	8.07

ZnO-C_3N_4 photoanode-based DSSCs and comparing them with pure ZnO-based DSSC. This study revealed that the coating of g-C_3N_4 improved the efficiency to 4.5% from 3.7% due to reduction in bandgaps and improvement in absorption spectrum. Also, improvement in injection of photogenerated electrons at photoanode and dye interface might be another factor for enhancement of the efficiency [15].

J. Xu et al. fabricated g-C_3N_4-modified TiO_2 nanosheets, prepared from different weight ratios (40, 60, 80, 100, and 150) of urea, as photoanodes using doctor blade method, and compared the performance. They found 80% wt. ratio modified nanosheets showed the highest efficiency with 7.34%, due to the creation of thin g-C_3N_4 over TiO_2 nanosheet surface, which retarded the backward recombination of electrons of TiO_2 and electrolyte and further enhanced the concentration of electrons in photoanodes [16]. Haoran Yan et al. fabricated solid-state DSSCs established on g-C_3N_4/TiO_2 and g-C_3N_4/Ag/TiO_2 nanocomposite photoanodes and compared with those based on pure TiO_2. They identified the addition of g-C_3N_4 has enriched the efficiency to 5.34% from 3.72% achieved by pure TiO_2 and further enhancement to 6.22% was observed after loading Ag to the composite. The optimal loading of Ag and g-C_3N_4 was identified to 2 and 5 wt.%, respectively. The addition of Ag and g-C_3N_4 not only improved the concentration of electrons but also retarded the backward recombination of electrons [19].

Z. Yuan et al. investigated the efficiency of DSSCs made with g-C_3N_4-modified TiO_2 nanosheets as photoanodes with Co_9S_8 nanoarrays and Pt as counter electrodes. The efficiency achieved by the DSSCs made with Co_9S_8 nanoarrays as counter electrodes (6.19%) was comparable to those fabricated with Pt as counter electrodes (6.79%). Further addition of g-C_3N_4 in TiO_2 photoanode enhanced the efficiency to 8.07% with Co_9S_8 nanoarrays as counter electrodes, because addition of g-C_3N_4 retards the recombination of electrons [28] (Table 1).

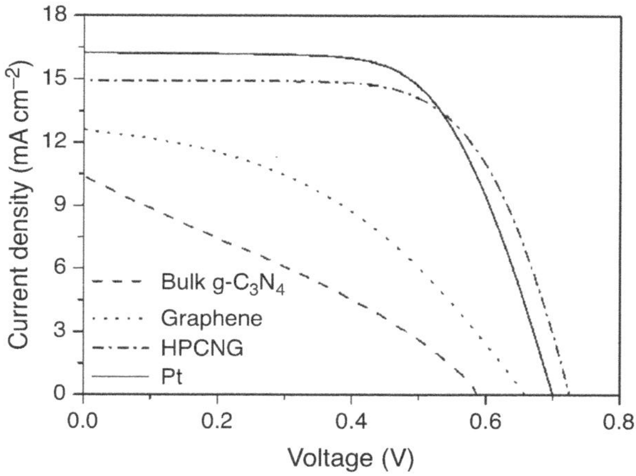

Fig. 6 J-V curves for DSSCs with different counter electrodes [20]

Table 2 I–V characteristics of DSSCs based on g-C_3N_4 and its composites as counter electrodes

References	DSSC photoanode	J_{sc} (mA/cm^2)	V_{oc} (Volts)	FF (%)	Efficiency (%)
Afshari et al.	PANI/g-C_3N_4 composite	7.56	0.59	–	1.786
Wang et al.	Bulk g-C_3N_4	10.26	0.587	0.31	1.87
	Graphene	12.59	0.659	0.42	3.49
	g-C_3N_4/graphene composite	14.91	0.723	0.66	7.13

3.2 *Cathode (Counter Electrode)*

Mohaddeseh Afshari et al. investigated the catalytic activity of polyaniline/graphitic carbon nitride (PANI/g-C_3N_4) nanocomposites, prepared by in situ electrochemical polymerization under sonication, as a counter electrode, in order to find the replacement to Pt electrode in DSSCs. They achieved the efficiency of 1.79%, and they found the composite has shown better catalytic activity in elevating the tri-iodide reaction, because of the synergetic effect of PANI and g-C_3N_4 [3]. Guiqiang Wang et al. fabricated DSSCs with graphene, g-C_3N_4, and g-C_3N_4/graphene nanocomposite as counter electrodes. By comparing, they identified that the efficiency of the DSSC with g-C_3N_4/graphene nanocomposite as a counter electrode showed a value of 7.13% which is higher than others and comparable to those with Pt as counter electrode, whose efficiency was found to be 7.37%. The enhancement by composite is due to large surface area, porous structure, and efficient electron transportation [22]. The J-V curves were represented in Fig. 6 (Table 2).

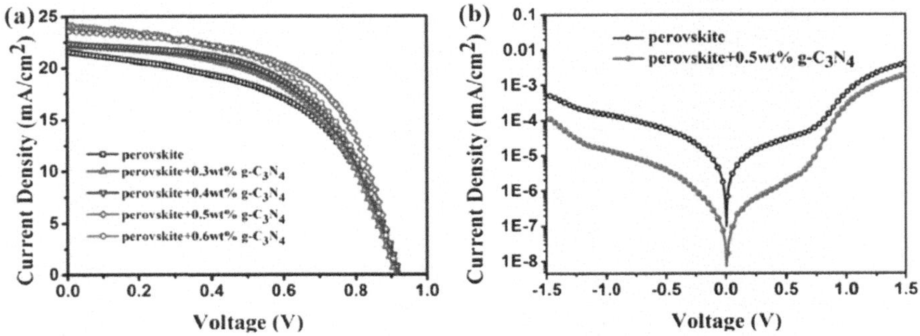

Fig. 7 (**a**) J-V measurements of the g-C_3N_4-doped PSC devices with selected ratios of precursor solutions. (**b**) The dark J-V measurements without and with g-C_3N_4 additive of the PSC devices plotted on a semilog scale [31]

4 Graphitic Carbon Nitrides on Perovskite Solar Application

The performance characteristics of the g-C_3N_4 mixed carbon-based perovskite solar cells (C-PSCs) are represented in Fig. 7a. The C-PSCs with the pure perovskite material (0 wt.% g-C_3N_4) reveal performance degradation with 0.925 V of V_{oc}, 21.5 mA/cm^2 of J_{sc}, and 0.529 FF and with the efficiency of 10.50% [31]. When the addition of a minor quantity of g-C_3N_4 in the perovskite material solution, which improves the V_{oc}, J_{sc}, and FF of the cell consistently with the efficiency higher than 12%, is measured due to the improved quality in perovskite layer. In precise, the C-PSCs with 0.5 wt % addition of g-C_3N_4 attain a higher efficiency of 12.85% [31]. For the g-C_3N_4-integrated device, the incident photon-to-current conversion efficiency (IPCE) is enriched, specifically because of the tiny wavelength region and significant improvement in the capability of light harvesting and hence results with more J_{sc}. Because of the dopant precipitation in the formation of C-PSC, the FF is decreased for the 0.6 wt.% addition of g-C_3N_4 [31]. Moreover, Fig. 7b represents the dark J-V characteristics of the C-PSC doped with g-C_3N_4 display less leakage current compared to that of pristine perovskite, signifying a larger resistance. The device with 0.5 wt.% addition of g-C_3N_4 resulted in 12.6% of steady-state efficiency. The photon-to-current conversion efficiency (PCE) and J_{sc} remain the same at 12.6% and 21.46 mA/cm^2 because of the integration of harmless g-C_3N_4 in the device.

In the C-PSCs, the lower V_{OC} initiated by the continuous carrier recombination are improved by developing a thin Al_2O_3 insulating layer over the ETM, m-TiO_2 with simple spin-coating method [31]. The physical barrier among ETM and perovskite alters the interface contact and conquers the recombination of charge. On the heat-treated (at 500 °C) substrates, the aluminum acetylacetonate in acetylacetone was spin-coated leading to the formation of Al_2O_3 layer as revealed in Fig. 8a. The g-C_3N_4-doped perovskite material is spin-coated over the Al_2O_3 layer. Figure 8b

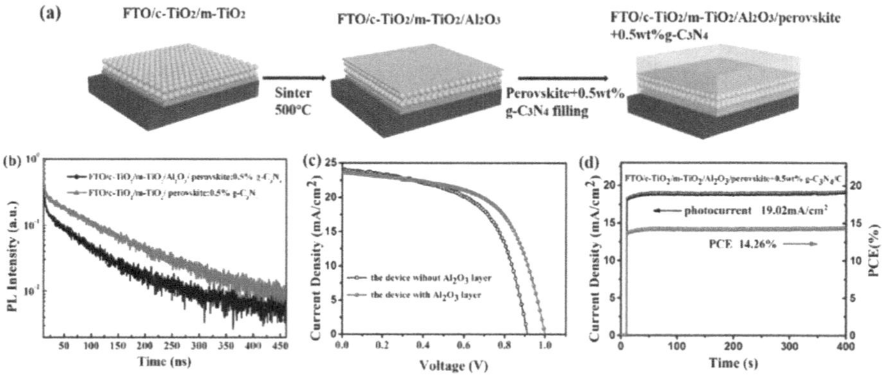

Fig. 8 (**a**) Spin coating of the Al_2O_3 insulating interlayer. (**b**) Time-resolved PL decay analysis of FTO/c-TiO_2/m-TiO_2/Al_2O_3/perovskite:0.5%g-C_3N_4 and FTO/c-TiO_2/m-TiO_2/perovskite:0.5%g-C_3N_4 layers (**c**) J-V measurements of PSC devices with and without Al_2O_3 insulating interlayer. (**d**) The higher steady-state output photocurrent at the maximal power position for the champion cell at 0.7 V [31]

reveals the excitons are highly active in c-TiO_2/m-TiO_2/Al_2O_3/perovskite:0.5%g-C_3N_4 with perovskite as an ETM than the c-TiO_2/m-TiO_2/perovskite:0.5%g-C_3N_4 with perovskite as an ETM. Subsequently, the insulating layer efficiently hinders the interface layer recombination and facilitates fast charge transport. In addition to the performance measurement of the device with the role of the Al_2O_3 (TiO_2/Al_2O_3/perovskite/carbon), the impedance spectra (IS) are measured for the cells without and with Al_2O_3 in dark mode. The Rs of the devices (TiO_2/Al_2O_3/perovskite/carbon) with Al_2O_3 (83 Ω) is coined as marginally greater than that of the device without Al_2O_3 (70 Ω) and indicates the insulating capability of physical barrier [31]. The agreement is remarkable compared to the recent literature. Nevertheless, the essential factor is recombination resistance (R_{rec}) in the device (TiO_2/Al_2O_3/perovskite/carbon) with Al_2O_3 (1576 Ω) significantly greater than that of the device without Al_2O_3 (1120 Ω); as additional components for both PSCs are comparable, the improved R_{rec} could replicate the consequence of the Al_2O_3 layer over conquering the recombination charge interface [31]. Since the improvement of R_{rec} is considerably greater than that of Rs, the ventured R_{rec} is the core parameter directing the process of charge transfer. Because of this, the charge transfer is enhanced. In addition to the energy level orientation in the C-PSCs (with/No insulating layer Al_2O_3), the electrons transport towards TiO_2 (4.0 eV) from perovskite (3.9 eV) and holes are brought together in carbon layer (5.0 eV) from perovskite (5.4 eV). The Al_2O_3 interlayer conduction band is directly over the TiO_2 conduction band, ensuing to inhibit electrons from TiO_2 transferring to the perovskite, which proceeds the recombination. In the meantime, the light insulating layer might alter the surface of ETM and be favorable for the charge transport. Accordingly, the thin Al_2O_3 enhanced the charge transport [31].

To relate the effect of Al_2O_3 insulating layer toward the C-PSCs performance measurement, FTO/c-TiO_2/m-TiO_2/perovskite+0.5wt%g-C_3N_4/C and FTO/c-TiO_2/m-TiO_2/Al_2O_3/perovskite+0.5wt%g-C_3N_4/C are considered, and the devices are discussed with and without Al_2O_3 C-PSCs. Figure 8c represents the J-V measurements of the C-PSCs by the Al_2O_3 layer are confirmed with standard illumination AM 1.5 G (100 mW/cm^2). The Al_2O_3 layer initiation considerably improves the V_{oc} and shows an increase of the efficiency in the fabricated devices. The C-PSCs (5000 rpm) have the maximum efficiency of 14.34% with V_{oc}, J_{sc}, and FF of 1.0 V, 23.86 mA/cm^{-2}, and 0.601 correspondingly as shown in Fig. 8c. Besides, the higher power point (0.7 V) of steady-state photocurrent output is illustrated in Fig. 8d. The illumination current density extends to the extremely very fast, and the attenuation is restrained for 400 s with a higher as 14.26% of the PCE [31]. The pure and optimized cells revealed a greater repeatability after the introduction of g-C_3N_4 and Al_2O_3. The reliability of the C-PSC is also evaluated in the present work, and the measure of uncapped cells (FTO/c-TiO_2/m-TiO_2/Al_2O_3/perovskite+0.5 wt%g-C_3N_4/C) is measured in the open-air environment [31]. After storage of 15 days, the intended device shows decent stability and extended 85% of higher PCE. This attribute to the interlayer of carbon electrode and the Al_2O_3 insulating layer can be ideal to secure the perovskite layer from the water vapor of the open air [31].

4.1 Influence of g-C_3N_4 Bulk or C_3N_4 Quantum Dots Structural Modification on the Measure of Inverted BHJ-Perovskite Solar Cells

The substance g-C_3N_4 or C_3N_4 quantum dots are modified with the blended solution of PTB7-Th:PC71BM, P3HT:PC61BM, or PBDTTT-C:PC71BM in *o*-dichlorobenzene by 0.2 mg.mL^{-1} as doping ratio. With the use of C_3N_4-modified P3HT:PC61BM mixture solution that acts photoactive layer, under air atmosphere ITO/ZnO/C_3N_4:P3HT:PC61BM/PEDOT:PSS/Ag reversed BHJ-perovskite cells are formed (Fig. 9a). From Fig. 9b, the J-V arcs for the substance C_3N_4 QDs or g-C_3N_4-modified P3HT:PC61BM cells are differentiated, which comprises the model (undoped) device. The constraints studied (cell parameters) are tabulated in Table 3. The reference cell (cell A) displays a 0.60 V as V_{oc}, a 10.09 mA.cm^{-2} J_{sc}, and a 58.2% FF and results a 3.60% (average value) as a PCE and equivalent to the previous work of the reversed BHJ-perovskite cells formed under the same environment (solvent as *o*-dichlorobenzene) [14]. As the substance g-C_3N_4 modified (cell B), the PCE displays a small alteration from 3.60% to 3.58% by a less similar photovoltaic property signifying that substance g-C_3N_4-modified P3HT:PC61BM active layer influences least in the device measurement. Nevertheless, the PCE in the device modified by C_3N_4 quantum dots (cell C) upsurges intensely as 4.23% computed for V_{oc} the values 0.61 V, J_{sc} as 11.44 mA.cm^{-2}, and FF as 60.2%. In the photovoltaic characteristics, the upsurge of J_{sc} about 13.4% enrichment is mainly

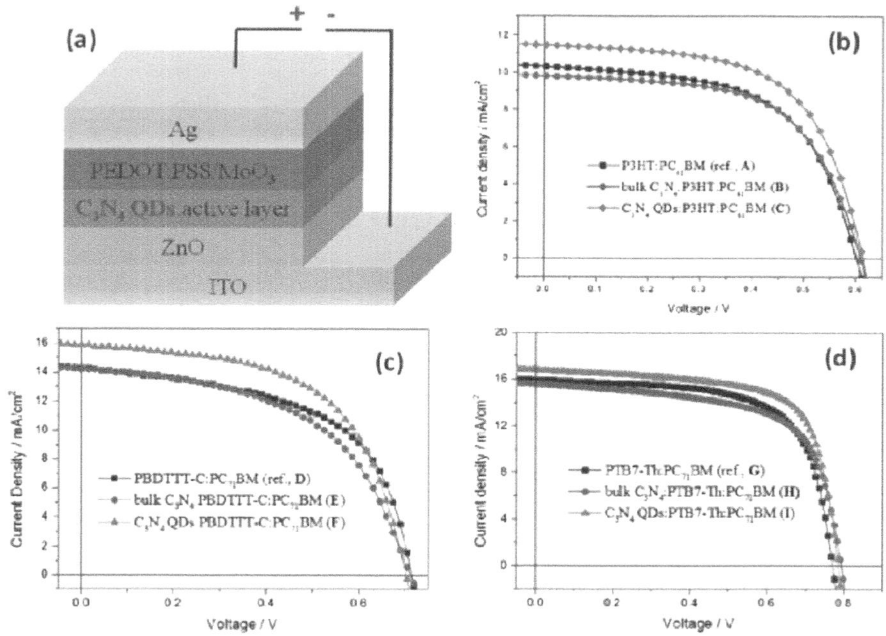

Fig. 9 (**a**) Device constructions ITO/ZnO/C_3N_4: active layer/PEDOT:PSS (or MoO_3)/Ag reversed BHJ-PSC schematic. (**b**) *J-V* arcs for pure and raw g-C_3N_4 or C_3N_4 QD-modified ITO/ZnO/C_3N_4: P3HT:PC61BM/PEDOT:PSS/Ag. (**c**) ITO/ZnO/C_3N_4:PBDTTT-C:PC71BM/MoO_3/Ag and (**d**) ITO/ZnO/C_3N_4:PBT7-Th:PC71BM/MoO_3/Ag devices analyzed by illuminating an AM 1.5 (100 mW.cm^{-2}) solar simulator in air [14]

liable for the PCE improvement of about 17.5% compared with the reference device, while V_{oc} and FF display only slight raise. The PCDTBT:PC71BM and P3HT: PC61BM doped with the carbon and CdS quantum dots, respectively, showed noticeable PCE improvements due to the upsurge in both J_{sc} and FF [14]. Remarkable, the seemingly dissimilar current incident of C_3N_4 QDs modification signifies the uniqueness of C_3N_4 QDs.

The usefulness of C_3N_4 quantum dots in improving the efficiency of P3HT: PC61BM reversed BHJ-perovskite cells inspired by examining its consequence of other high efficient cells established under low-bandgap donor substantial materials with PBDTTT-C and PTB7-Th mixed PC71BM is described in recent times for the PCE about 7.2% and 10.0%, correspondingly [14]. Figure 9c displays the *J-V* arcs for the ITO/ZnO/C_3N_4:PBDTTT-C:PC71BM/MoO_3/Ag reversed BHJ-perovskite cells made up using glove box (shown in Fig. 8a) as compared with the orientation (undoped) cell. The orientation cell (cell D) displays a 5.70% PCE. Once the C_3N_4 quantum dots are doped, PCE of the cell (cell F) rises intensely to 6.36%. Remarkably, compared with the P3HT:PC61BM structure, PBDTTT-C:PC71BM structure shows improvement in PCE of about 11.6% absolute as that of the substance

Table 3 Photovoltaic characteristics of the ITO/ZnO/C_3N_4:PBDTTT-C:PC71BM/MoO_3/Ag devices of various activation layers [14]

Active layer	Device	Doping	Voc (V)	Jsc (mA/cm^{-2})	FF (%)	PCE (%) Highest	PCE (%) Average
P3HT:PC61	A	–	0.60	10.09	59.8	3.80	3.60 ± 0.15
	B	Bulk C_3N_4	0.61	9.80	59.9	3.75	3.58 ± 0.13
	C	C_3N_4 QDs	0.61	11.44	60.2	4.56	4.23 ± 0.25
PBDTTT-C: PC71BM	D	–	0.72	14.30	55.4	5.83	5.70 ± 0.20
	E	Bulk C_3N_4	0.72	14.24	51.5	5.50	5.28 ± 0.09
	F	C_3N_4 QDs	0.70	15.90	57.1	6.62	6.36 ± 0.21
PTB7-Th: PC71BM	G	–	0.77	15.91	67.2	8.39	8.21 ± 0.15
	H	Bulk C_3N_4	0.79	15.56	64.6	7.96	7.86 ± 0.22
	I	C_3N_4 QDs	0.78	16.74	69.9	9.20	9.18 ± 0.19

structure created mainly by the raise of J_{sc} (to 15.90 mA.cm^{-2} from 14.30 mA.cm^{-2}, about 11.2% improvement), while minor fluctuations (tradeoff) are perceived for V_{oc} and FF. Alternatively, the cell structure by bulk g-C_3N_4 modification revealed a slight declined PCE (5.50%) as a substitute because of the reduction in FF. By the use of cell with PTB7-Th:PC71BM activation layer, conferring as the evaluation for the *J-V* arcs without or with C_3N_4 quantum dots modification presented in Fig. 9d, the PCE in the C_3N_4 quantum dots modified cell structure (cell I) computed with V_{oc} as 0.78 V, a J_{sc} as 16.74 mA.cm^{-2}, and 69.9% of FF as stretches 9.18%, which is improved by about 11.8% associated with the substance scheme (cell G) signifying an 8.21% PCE. On the other hand, raw C_3N_4 modification in PTB7-Th:PC71BM activation layer (cell H) displays a reduced PCE as 7.86%. The present occurrences are pretty comparable for the P3HT:PC61BM structure, authorizing that the development in C_3N_4 quantum dots is essential on the efficiency enrichment process [14]. Remarkably, by the usage of PBDTTT-C:PC71BM structure, the J_{sc} (about 11.2%) increases to donate mainly for the PCE enrichment (about 11.6%), while for PTB7-Th:PC71BM structure, the J_{sc} (about 5.2%) increase donates a smaller amount of the PCE development (about 11.8%), and FF rises about 4.0% concurrently. For the two structures PBDTTT-C:PC71BM and PTB7-Th:PC71BM, J_{sc} increases upon C_3N_4 quantum dots modification, which is reliable in the EQE effect and displays a noticeable raise of EQE in the complete range of 340–720 nm (as PBDTTT-C:PC71BM) or 300–680 nm (as PTB7-Th:PC71BM). In the note, nevertheless in the structure PTB7-Th:PC71BM, which is on the specific range as 420–510 nm and the EQE for C_3N_4 quantum dots modified structure is lesser than that of undoped device. This is because of the reduction in absorption of light from the same region. Instead, for the two devices PBDTTT-C:PC71BM and PTB7-Th:

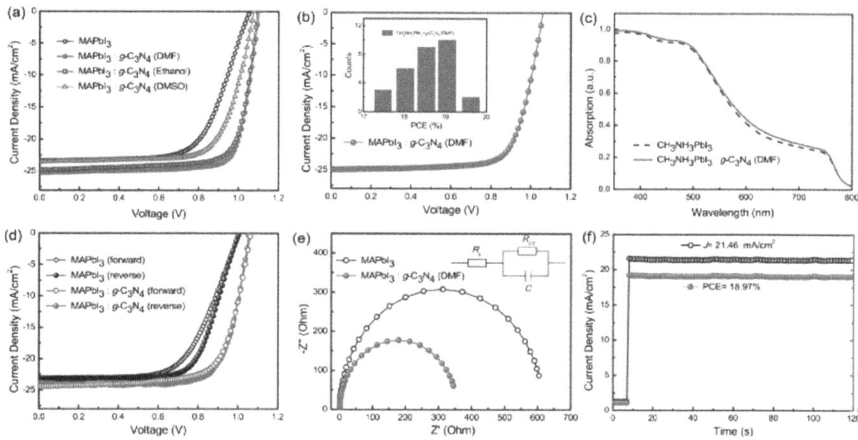

Fig. 10 (**a**) *J-V* arcs analyzed with simulated AM 1.5 sunlight of 100 $mWcm^{-2}$ for $CH_3NH_3PbI_3$- and $CH_3NH_3PbI_3$:g-C_3N_4-constructed perovskite solar cells. (**b**) PCE histogram analyzed from 30 devices constructed with $CH_3NH_3PbI_3$:g-C_3N_4. (**c**) UV-Vis spectra for $CH_3NH_3PbI_3$ and $CH_3NH_3PbI_3$:g-C_3N_4 (DMF) substance coated over TiO2/FTO. (**d**) *J-V* arcs measured under forward and reverse scan of $CH_3NH_3PbI_3$- and $CH_3NH_3PbI_3$:g-C_3N_4-constructed devices. (**e**) Nyquist plots performed under the dark field and on an open voltage for devices constructed with $CH_3NH_3PbI_3$ and $CH_3NH_3PbI_3$:g-C_3N_4. (**f**) Maximal steady-state output photocurrent for $CH_3NH_3PbI_3$:g-C_3N_4-constructed device and their respective power output at the maximum power point of 0.95 V [32]

PC71BM, the total EQE of raw g-C_3N_4_modified structures are lesser when compared with undoped device structures, assenting to the decline of J_{sc} value [14].

4.2 Photovoltaic Performance

The *J-V* characteristic analysis for $CH_3NH_3PbI_3$ (without and with *g*- C_3N_4)-structured PSCs in AM 1.5G radiance with 100 mW.cm^{-2} of light intensity is shown in Fig. 10a. The orientation device provides 16.22% of PCE with 23.33 mA.cm^{-2} as J_{sc}, 1.06 V as V_{oc}, and 0.65 as FF. With the inclusion of *g*-C_3N_4 material in the perovskite layer, the efficiency of the device increases irrespective of the solvent used for addition of *g*-C_3N_4. Predominantly, the $CH_3NH_3PbI_3$:*g*-C_3N_4 (DMF)-coated device reveals maximum 19.49% as PCE with 24.31 mA.cm^{-2} as J_{sc}, 1.07 V as V_{oc}, and 0.74 as FF [32]. A PSC constructed with sequence of $CH_3NH_3PbI_3$:*g*-C_3N_4 (DMF) is established for the performance reproducibility confirmation. Figure 10b intrigues the consistent PCEs histogram studied from more than 30 devices. The peak performance of PSCs based on *g*-C_3N_4 (DMF) exhibits maximum mean PCE (PCEAVE) of 19.14%. Figure 10c represents the optical absorption arcs of coated layers made of $CH_3NH_3PbI_3$ and $CH_3NH_3PbI_3$:*g*-C_3N_4 over TiO_2/FTO substance. A minor improvement is detected from the entire visible spectral series through g-C_3N_4 alteration, and the same is reliable by the slight modification of the J_{sc} in resultant PSCs. The *J-V* hysteresis actions repeatedly look like utmost high-performance PSCs [32]. Moreover, the influence in ion relocation and the shortcomings of bulk perovskite layer and on the boundary are besides observed as the prime factor of the hysteresis performance [32]. As abovementioned, g-C_3N_4 plays a high significant part in coating the perovskite layer with half-reduced imperfections on current instance. The better layer coverage in g-C_3N_4-composed perovskite film recommends a condensed carrier recombination of the grain boundaries because of the removed trap modes [32]. Essentially, the hysteresis of current is intensely reliant in the interaction states among perovskite film and interfacial layers. The surface modification of perovskite film is decreased by means of g-C_3N_4 passivation, causing a better edge interaction among $CH_3NH_3PbI_3$ and Spiro-OMeTAD layers [32]. These enhancements make the *J-V* arcs as a slight hysteresis in devices integrated with g-C_3N_4. Consequently, PSCs composed of g-C_3N_4 are predictable in display compact *J-V* hysteresis performance [32]. Figure 10d displays the *J-V* arcs of substance and PSCs made of $CH_3NH_3PbI_3$: g-C_3N_4 through forward and reverse sweep information. The substance device displays a great hysteresis of current associated with huge alteration among the 15.43% PCEs during the forward scan and 15.92% from reverse scan [32]. In similarity, the PSC constructed with $CH_3NH_3PbI_3$:g-C_3N_4 offered a reduction in hysteresis of current by exactly 19.02% PCEs during the forward scan, and in reverse scan it is from 19.03%. Prominently, the enhancement in PCEs of all PSCs incorporated with g-C_3N_4 is largely ascribed to the improved FF from 0.65 to 0.74. The passivation of g-C_3N_4 through the flat perovskite layer and better edge interaction with the upper layer [32]. The influence of g-C_3N_4 in the cell constancy, the efficiency, and steady-state photocurrent are dignified in the extreme power point (0.95 V) as revealed in Fig. 10f [32]. The current density and the PCE are alleviated by 21.46 $mA.cm^{-2}$ and 18.97%, correspondingly [32]. The indications show that the integration of g-C_3N_4 cannot depreciate the stability of the cell extremely. On all, g-C_3N_4 integration plays a major passivation consequence on perovskite film, subsequent of an apparent development in the PCE impending 19.49% [32].

5 Summary and Conclusion

In this chapter, we have been presented the different synthesis approaches for the preparation of graphitic carbon nitrides (g-C_3N_4) for dye-sensitized solar cells (DSSCs) and perovskite solar cells. First, the material preparation and structural properties were investigated. An ultrasonic-assisted liquid-phase exfoliation and solvothermal method were used for the formation of g-C_3N_4 thin film. Next, the structural properties of nature of the materials, unit cell parameters, and crystallite size (Debye-Scherrer) are calculated using X-ray diffractometer techniques. Further, the prepared thin film morphological investigation was carried out using scanning electron microscopy and transmission electron microscopy. The g-C_3N_4 is used in

DSSC devices as photoanodes, counter electrodes, and doping. The cell efficiency of g-C_3N_4 in TiO_2 photoanodes has achieved the maximum 8.07% with Co_9S_8 nanoarrays as counter electrodes. Next, g-C_3N_4-based nanocomposites used as a counter electrode (polyaniline/g-C_3N_4) reached up to 1.79% of cell efficiency. However, g-C_3N_4/graphene composite was used as a counter electrode and significantly yielded 7.13% of conversion efficiency due to the large surface area and porous structure and efficient electron transport within the device. Furthermore, extended this material (g-C_3N_4) performance was investigated in perovskite solar cells (PSCs) by doping. In PSC, 0.5 wt% of g-C_3N_4 yielded 12.85% of cell efficiency and 24 mA/cm^2 of short-circuit current. This promising device attained higher cell performance at shorter-wavelength regions and improved light harvesting mechanisms. The additional improvement was noticed by integrating the insulator thin film layer between the PSC devices. Here, Al_2O_3 consists of an insulating layer which helped to improve the photovoltaic performance and remarkably gained 14.34% and 23.86 mA/cm^2 of cell efficiency and short-circuit current. The g-C_3N_4 and C_3N_4 quantum dots were doped with P3HT:PC61, PBDTTT-C:PC 71 BM. By adding C_3N_4-doped P3HT:PC61BM mixture solution used as the photoactive layer, ITO/ZnO/C_3N_4:P3HT:PC 61 BM/PEDOT:PSS/Ag inverted BHJ-PSC devices are formed and reached efficiency of 3.6%. Similarly, the devices with active layer of PTB7-Th:PC71BM modified by addition of C_3N_4 quantum dots showed 11.8% of conversion efficiency. This doped C_3N_4 quantum dot was reliable with the external quantum effect and noticed the enhanced absorption from 340 to 720 nm spectral region. Continuously, this PSC inclusion of the g-C_3N_4 materials layer improves the cell performance with the effect of various combinations of nanostructures such as active layer, photoanode, counter electrode, doping, etc. Predominantly, $CH_3NH_3PbI_3$:g-C_3N_4 (DMF) cell structure significantly enhanced the remarkable photovoltaic conversion efficiency of 19.49% with short-circuit current of 24.31 mA/cm^2. In summary, the presented DSSC and perovskite solar cells would be helpful to the new generation of g-C_3N_4 thin film and quantum dot-based solar cells.

Acknowledgments Authors acknowledge the support of DST-SERB India under Core Research Grant (CRG/2019/005985).

References

1. Fan K, Yu J, Ho W (2017) Improving photoanodes to obtain highly efficient dye-sensitized solar cells: a brief review. Mater Horizons 4(3):319–344
2. Li C, Cao Q, Wang F, Xiao Y, Li Y, Delaunay JJ, Zhu H (2018) Engineering graphene and TMDs based van der Waals heterostructures for photovoltaic and photoelectrochemical solar energy conversion. Chem Soc Rev 47(13):4981–5037
3. Afshari M, Dinari M, Momeni MM (2018) Ultrasonic irradiation preparation of graphitic-C_3N_4/polyaniline nanocomposites as counter electrodes for dye-sensitized solar cells. Ultrason Sonochem 42:631–639

4. Lv H, Yuan X, Cui C (2017, March) Enhancing the photovoltaic performance of dye-sensitized solar cells by modifying TiO_2 photoanodes with layered structure g-C_3N_4. In: AIP conference proceedings, vol. 1820, no. 1. AIP Publishing LLC, p 040010
5. Venumbaka MR, Akkala N, Duraisamy S, Sigamani S, Poola PK, Marepally BC (2022) Performance of TiO_2, Cu-TiO_2, and N-TiO_2 nanoparticles sensitization with natural dyes for dye sensitized solar cells. Mater Today Proc 49:2747–2751
6. Selvakumar D, Murugadoss G, Alsalme A, Alkathiri AM, Jayavel R (2018) Heteroatom doped reduced graphene oxide paper for large area perovskite solar cells. Sol Energy 163:564–569
7. Selvakumar D, Vasudevan R, Jayavel R (2018) Formation of PbSe–ZnO thin film based heterostructure for solar cell applications. Mater Today Proc 5(6):14468–14472
8. Halevy S, Korin E, Bettelheim A (2016) Electropolymerization as a new route to g-C_3N_4 coatings on TiO_2 nanotubes for solar applications. RSC Adv 6(90):87314–87318
9. Ansari MS, Banik A, Qureshi M (2017) Morphological tuning of photo-booster g-C_3N_4 with higher surface area and better charge transfers for enhanced power conversion efficiency of quantum dot sensitized solar cells. Carbon 121:90–105
10. Yang X, Zhao L, Wang S, Li J, Chi B (2021) Recent progress of g-C_3N_4 applied in solar cells. J Mater 7(4):728–741
11. Ai L, Shi R, Yang J, Zhang K, Zhang T, Lu S (2021) Efficient combination of G-C_3N_4 and CDs for enhanced photocatalytic performance: a review of synthesis, strategies, and applications. Small 17(48):2007523
12. Alaghmandfard A, Ghandi K (2022) A comprehensive review of graphitic carbon nitride (g-C_3N_4)–metal oxide-based nanocomposites: potential for photocatalysis and sensing. Nano 12(2):294
13. Reddy KR, Reddy CV, Nadagouda MN, Shetti NP, Jaesool S, Aminabhavi TM (2019) Polymeric graphitic carbon nitride (g-C_3N_4)-based semiconducting nanostructured materials: synthesis methods, properties and photocatalytic applications. J Environ Manag 238:25–40
14. Chen X, Liu Q, Wu Q, Du P, Zhu J, Dai S, Yang S (2016) Incorporating graphitic carbon nitride (g-C_3N_4) quantum dots into bulk-heterojunction polymer solar cells leads to efficiency enhancement. Adv Funct Mater 26(11):1719–1728
15. Wu D, Cao K, Wang F, Wang H, Gao Z, Xu F, Guo Y, Jiang K (2015) Two dimensional graphitic-phase C_3N_4 as multifunctional protecting layer for enhanced short-circuit photocurrent in ZnO based dye-sensitized solar cells. Chem Eng J 280:441–447
16. Xu J, Wang G, Fan J, Liu B, Cao S, Yu J (2015) G- C_3N_4 modified TiO_2 nanosheets with enhanced photoelectric conversion efficiency in dye-sensitized solar cells. J Power Sources 274: 77–84
17. Li X, Pan K, Qu Y, Wang G (2018) One-dimension carbon self-doping g- C_3N_4 nanotubes: synthesis and application in dye-sensitized solar cells. Nano Res 11(3):1322–1330
18. Yan H, Tian X, Pang Y, Feng B, Duan K, Zhou Z, Weng J, Wang J (2016) Heterostructured g-C_3N_4/Ag/TiO_2 nanocomposites for enhancing the photoelectric conversion efficiency of spiro-OMeTAD-based solid-state dye-sensitized solar cells. RSC Adv 6(104):102444–102452
19. Gkini K, Martinaiou I, Falaras P (2021) A review on emerging efficient and stable perovskite solar cells based on g-C3N4 nanostructures. Materials 14(7):1679
20. Wang G, Zhang J, Kuang S, Zhang W (2016) Enhanced electrocatalytic performance of a porous G-C_3N_4/graphene composite as a counter electrode for dye-sensitized solar cells. Chemistry 22(33):11763–11769
21. Safaei J, Mohamed NA, Noh MFM, Soh MF, Ludin NA, Ibrahim MA, Isahak WNRW, Teridi MAM (2018) Graphitic carbon nitride (g-C_3N_4) electrodes for energy conversion and storage: a review on photoelectrochemical water splitting, solar cells and supercapacitors. J Mater Chem A 6(45):22346–22380
22. Jia L, Cheng X, Wang X, Cai H, He P, Ma J, Li L, Ding Y, Fan X (2019) Large-scale preparation of g-C_3N_4 porous nanotubes with enhanced photocatalytic activity by using salicylic acid and melamine. Ind Eng Chem Res 59(3):1065–1072

23. Liu J, Zhang T, Wang Z, Dawson G, Chen W (2011) Simple pyrolysis of urea into graphitic carbon nitride with recyclable adsorption and photocatalytic activity. J Mater Chem 21(38): 14398–14401
24. Amorin LH, Suzuki VY, de Paula NH, Duarte JL, da Silva MAT, Taft CA, La Porta FDA (2019) Electronic, structural, optical, and photocatalytic properties of graphitic carbon nitride. New J Chem 43(34):13647–13653
25. Thomas A, Fischer A, Goettmann F, Antonietti M, Müller JO, Schlögl R, Carlsson JM (2008) Graphitic carbon nitride materials: variation of structure and morphology and their use as metal-free catalysts. J Mater Chem 18(41):4893–4908
26. Yuan Z, Tang R, Zhang Y, Yin L (2017) Enhanced photovoltaic performance of dye-sensitized solar cells based on Co9S8 nanotube array counter electrode and TiO_2/g-C_3N_4 heterostructure nanosheet photoanode. J Alloys Compd 691:983–991
27. Qiu P, Chen H, Xu C, Zhou N, Jiang F, Wang X, Fu Y (2015) Fabrication of an exfoliated graphitic carbon nitride as a highly active visible light photocatalyst. J Mater Chem A 3(48): 24237–24244
28. Zhu J, Xiao P, Li H, Carabineiro SA (2014) Graphitic carbon nitride: synthesis, properties, and applications in catalysis. ACS Appl Mater Interfaces 6(19):16449–16465
29. Reddy VM, Rao DS (2017) Fabrication and performance evaluation of natural dye sensitized solar cells using TiO_2 incorporated with Ag nanoparticles. Int Res J Eng Technol 4(7): 2325–2327
30. Reddy VM, Lavanya K (2018) A study on the effect of silver nano colloidal solution on the efficiency of dye sensitized solar cell (DSSC). Int J Adv Innov Res 5(4):6–9
31. Yang ZL, Zhang ZY, Fan WL, Hu CS, Zhang L, Qi JJ (2019) High-performance g-C_3N_4 added carbon-based perovskite solar cells insulated by Al2O3 layer. Sol Energy 193:859–865
32. Jiang LL, Wang ZK, Li M, Zhang CC, Ye QQ, Hu KH, Lu DZ, Fang PF, Liao LS (2018) Passivated perovskite crystallization via g-C_3N_4 for high-performance solar cells. Adv Funct Mater 28(7):1705875
33. Esgin H, Caglar Y, Caglar M (2022) Photovoltaic performance and physical characterization of Cu doped ZnO nanopowders as photoanode for DSSC. J Alloys Compd 890:161848

Biomedical Devices Adopting Energy-Harvesting Schemes

M. Saravanan, Eswaran Parthasarathy, J. Ajayan, T. Shanmugaraja, J. Mercy, R. Lawanya, and B. Jaishankar

1 Introduction

In the 1950s, when advances in semiconductor technology allowed for entirely embedded pacemakers, doctors first began to consider the possibility of utilizing implanted medical devices to aid patients. With the intention of helping a patient with cardiovascular issues, the first human pacemaker was placed in 1958. The monitoring of blood pressure for hypertension, glucose levels for diabetes, and brain activity for a wide range of neurological diseases are other innovative and emerging scientific uses [1]. Embedding a patient for longer than they would normally live is awesome. The biosensor is shown as a square block with the outline of a battery underneath (Fig. 1).

Increasing battery capacity or reducing framework solidity improves embedded longevity [2]. Mild battery improvements, alternate fueling, and green hardware are review center issues. The embed must be under 1 cubic centimeter to fit inside the human shape and not disrupt normal activities. Biocompatibility prevents immune responses, rejection, and other problems. Battery-powered implants are common. A

M. Saravanan (✉)
Sri Eshwar College of Engineering, Coimbatore, Tamilnadu, India

E. Parthasarathy
SRM Institute of Science and Technology, Chennai, Tamilnadu, India

J. Ajayan
SR University, Warangal, Telangana, India

T. Shanmugaraja · B. Jaishankar
Department of ECE, KPR Institute of Engineering and Technology, Coimbatore, Tamilnadu, India

J. Mercy · R. Lawanya
Dr. N.G.P. Institute of Technology, Coimbatore, Tamilnadu, India

A. Nella et al. (eds.), *Energy Harvesting Trends for Low Power Compact Electronic Devices*, EAI/Springer Innovations in Communication and Computing,
https://doi.org/10.1007/978-3-031-35965-1_4

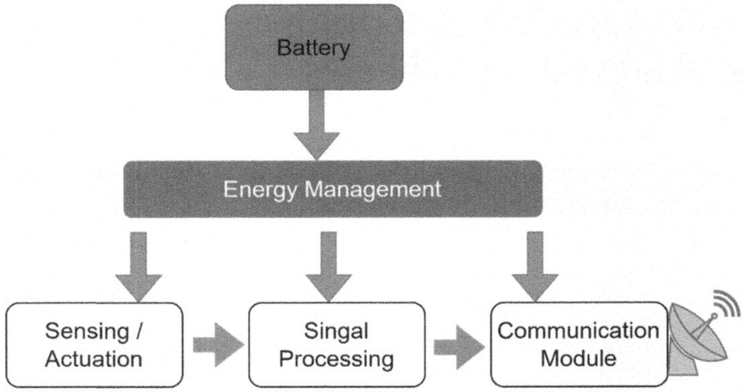

Fig. 1 Implantable biosensor implemented by battery

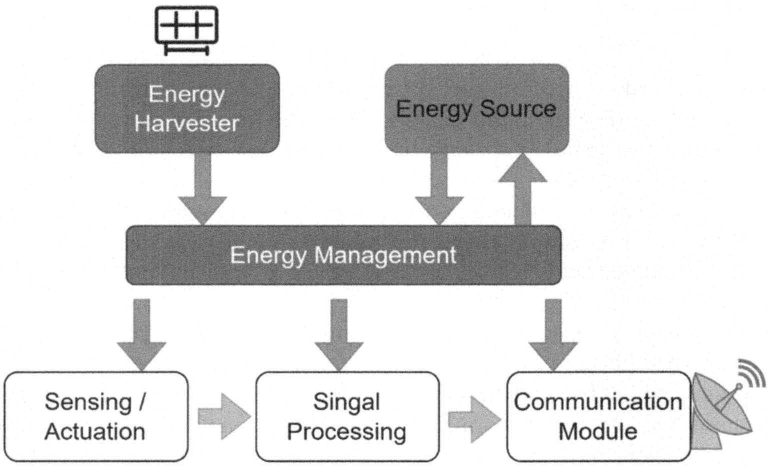

Fig. 2 Implantable biosensors powered by energy harvesting

battery's lifespan is limited. The battery will reduce the implant's size [2, 3]. Implantable devices may be hampered by battery limits. Increase implant presence without compromising structure to maintain growth. Energy harvesting is evident given the battery's limitations. Heat, vibrations, light, and chemical bonds can be converted to electric power. Energy harvesting can replace the battery, giving the gadget independence [4]. Power harvesting improves the battery by shrinking its size and extending its life. Fueling without batteries reduces battery issues.

Even if an implant is powered by power-gathering gear, it needs a battery. Figure 2 shows implantable biosensors powered by energy harvesting. A fully autonomous embed with energy-harvesting capabilities and a reliable electric-controlled capacity device would advance medical device development and IoMT. Power is unpredictable and scarce; thus even low amounts won't last long. The second one needs both an AC-to-DC and DC-to-DC voltage shift, but the first one simply needs a DC-to-DC shift and power-collection equipment. A single interface will regulate high skill levels, modify and up-convert voltage, and power implant electronics. The interface must also capture as much energy as possible from the source and deliver it to the power storage device and implant circuits. The interface's purpose is to send as little electricity as feasible [4, 5]. For implantable scientific instruments, energy-harvesting interfaces must handle low and rapid energy levels while operating better at greater passage energy levels.

2 Methods of Energy Harvesting and Its Sources

There are a wide variety of implantable biomedical devices on the market. The primary apparatus relies on the electrical current from the human body, and the supplementary techniques are all arranged around it. The second category consists of secondary forest-surrounded, renewable energy-powered communities. In biology, we employ every known method of capturing electrical energy (Fig. 3).

The motion of people and the heat from their bodies produce electricity. It is possible to generate power in a number of different ways by utilizing the body's massive physiological operations. In comparison to the 1630 mW produced by a full-out foot and action sprint, a sphere generates 81 mW of power while sleeping. Humans are capable of keeping their core temperature stable despite drastic shifts in ambient conditions [6]. The metabolic processes necessary for energy production can continue to function thanks to this quality, even when the temperature is quite

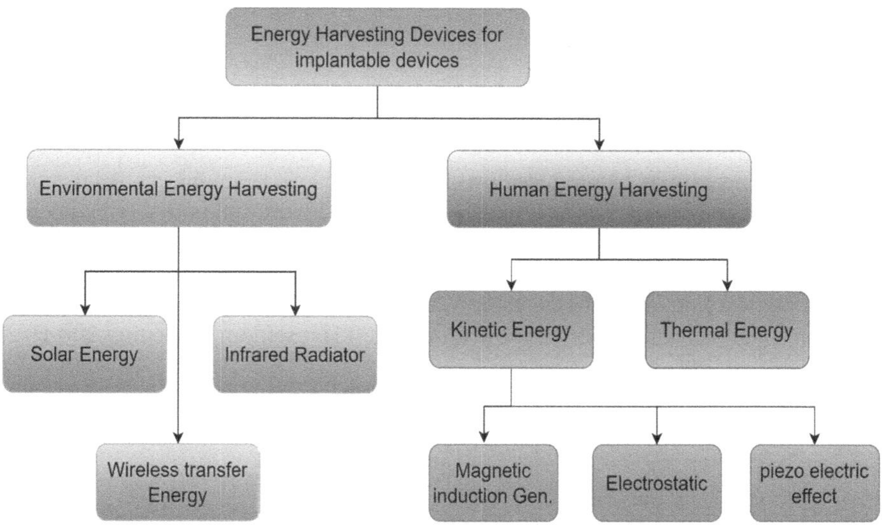

Fig. 3 Different kinds of energy-harvesting methodology

Fig. 4 Various sources of energy-harvesting methods

low. Therefore, the next section differentiates between kinetic energy harvesting and thermal energy harvesting, the two forms of power generated by human frame actions. Figure 4 depicts the various energy-gathering methods.

2.1 Mechanical Energy Harvesting

Mechanical energy may be produced by vibrations, shocks from objects like wet tree branches, tension switches on skis or skateboards, and worn-out or newly laundered clothing. With an electric-powered actuator (which includes a computer, microwaves, etc.), each at its constant frequency or a harmonic, or at its rotation speed, a vibration top can often be visible. When no strong emotions are occurring, vibration frequencies are normally low around 100 Hz, seldom exceeding 200 Hz.

2.2 *Kinetic or Vibration Energy*

Biological motions could power biomedical implants. Kinetic electricity is a convenient energy source [7]. Piezoelectric, magnetic induction generator, and electrostatic transducers are defined.

2.3 *Piezoelectric Effect*

Piezoelectricity converts mechanical to electrical energy. In 1990, MIT's Media Lab created the first energy-harvesting device powered by walking (MIT). With this invention, a person's movement-powered wearable digital devices. Paradiso et al. built a 1 W spring magnetic generator in 1998 [6, 7]. This energy-generating prototype is difficult to use and should only be used on patients who can walk.

There are two places in the shoe, the heel and the toes, where piezoelectric components are built in to make kinetic energy. Both 8.3 mW and 1.33 mW of power come from this prototype. On the other hand, this prototype can't generate as much power and would be best for people who can walk well. In the heel of a shoe, you'll find pressed piezoelectric materials. Figure 5 shows the whole process of how energy is changed. Even though the amount of power it produces has gone up, this prototype still has the same problems as the last one.

2.4 *Electrostatic Energy*

Electrostatic generators create electricity via electrostatic induction. These devices convert mechanical vibration to electric energy by moving a transducer against an electrical field. Constant charge and fixed voltage conversions. This works for low-power micro-implanted devices (along with implantable biosensors). Switched regular fee structures, switched regular voltage structures, and nonstop electret-based full structures [8]. Electret-loose capacitors can work in feed-limited and voltage-limited topologies. Electricity powered the devices in each case. Before using this way to create electricity, harvesters must charge the capacitor. Electronics with electrets are identical to those without. Adding electret layers to one or all capacitor plates improves polarization. Electrostatic harvesters increasingly use electrets. Even if they're made like magnets, it's difficult. It involves overheating

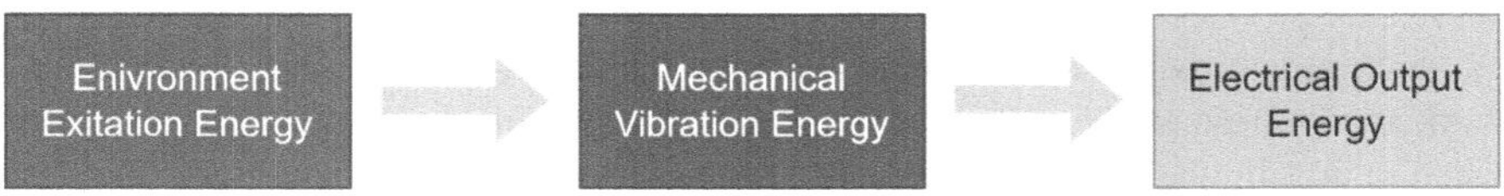

Fig. 5 Energy conversion flow

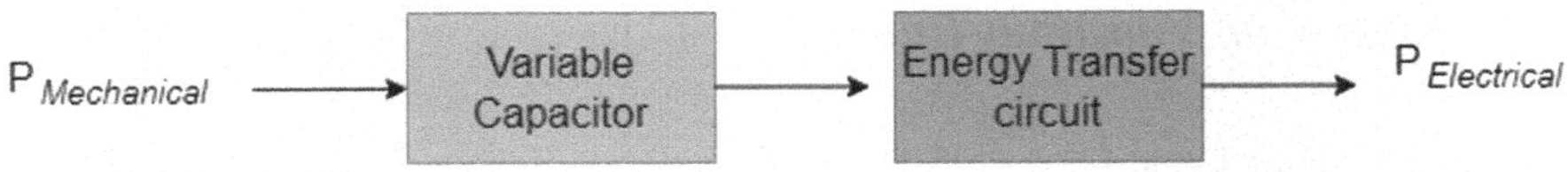

Fig. 6 Block of electrostatic energy harvester

or overcharging a dielectric layer. Cooling the layer maintains an electric field [7]. Electrostatic energy generators use variable capacitors. External pressure can move the plates, and the mills can run at a fixed price or voltage. External pressure reduces the capacitor's voltage, while plate movement reduces its current (Fig. 6).

2.5 Magnetic Induction Generator

Mechanical generators can produce electromagnetic energy. The first employs relative motion, where the generator stays static, while the second uses rigid frame motion, where the generator feels inertia pressure. By rotating the circuit around an axis and lowering the ground, electromagnetic transducers can display how magnetic flux changes. This technology previously powered a quartz watch. Heartbeats may be utilized in the future to rate pacemaker batteries using the "Seiko kinetic" method. Electromagnetic transducers convert mechanical force to electric force using Faraday's law [9]. Moving magnetic mass modifies magnetic flux in a neighboring coil. AC voltage results. Magnetic mill energy is inversely related to size. The number of coils turns and device size determines strength. Over time, friction from moving electromagnetic generators could hurt someone. This shortens harvesters' lives. Very small electromagnetic transducers can be made for low-frequency purposes.

3 Thermal Energy

The thermoelectric energy harvester makes electricity from minor temperature variations (Seebeck effect). Neurostimulators, cochlear hearing aids, and wireless patient diagnostics may demand this much power. A thermoelectric generator consists of a thermopile, which is a set of thermocouples connected in parallel. Thermoelectric energy harvesting depends on the generator's performance (TEG). A TEG swiftly converts heat into electricity via the Seebeck effect. The movement of electrons and holes generates a temperature differential [9, 10]. Thermoelectric harvesters produce green energy that can be stored. Silent and quiet; excellent for the environment because waste heat is captured and turned into energy; and able to work at very high temperatures. TEG devices need the most power and performance when in use. In waste heat therapy, only the electricity created is beneficial; the heat is squandered. Thermal energy harvesting only works 5% of the time, which is why it's not popular [10]. TEG performance has grown by 10% in recent years due to new thermoelectric materials.

3.1 Energy-Harvesting Sources in the Human Body

Sometimes it's tougher to power a biosensor with a moving human body than a vibrating commercial gadget. Human harvesters have physical restrictions and move less regularly, causing extra labor. Heat gradients between a person's body and the environment and inside the body are lower than in a commercial plant. This complicates thermoelectric harvesting. The amount of energy a person can give using large energy-gathering equipment, such as a shoe, is studied. An internal problem has worsened. Solar, thermal, and inertial kinetic energy-harvesting systems can be hermetically sealed, but fuel cells can't. Scalability could be vital for a biosensor's safety.

3.2 Thermal Energy Harvesting

Several technical solutions can be employed to build flexible, wearable electronic parts. Embroidery is the most promising approach to building conductive networks in fabric and clothing pieces. Connectors, resistive, and inductive additives can change the shape of garments while keeping their qualities. In recent years, embroidery has been recommended for making inductive circuit pieces. Thermoelectric generators can be used to store wearable power. These rigid harvesters are utilized for wristwatches. Thermoelectric components produced commercially can be used in clothing [11]. These harvesters make DC, which can be made by doing something odd or too much.

In this study, the notion of producing a power transformer by combining the movement of permanent magnets and flat spiral inductors is examined. We'll discuss how these mills are created, their features, how well they perform, and how they might be employed in clothes. Inductors are flattened using electricity harvesters. Human-shaped harvesters. Several well-known protocols show that square-shaped coils provide the maximum electricity (magnet length is thought to be the same to coil length). Experiments confirm theoretic predictions for different-sized and different-shaped inductors. Many approaches to producing a flat inductor include the fabric industry [11]. A flexible PCB is introduced. This approach can be utilized to produce precise inductors and power-controlling elements on a flexible substrate. Due to its lightness, the end product can be added to clothing without any alterations. The coil resistance is too high because copper rails are generally longer and thinner than 100 m so they can be twisted.

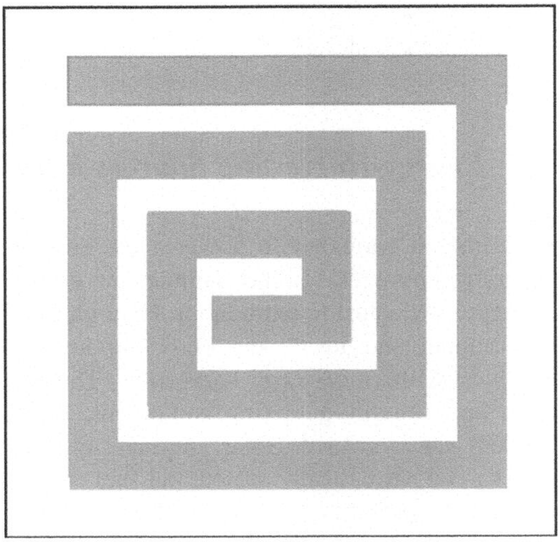

Fig. 7 Flat square-shaped inductor coil

A longer conductor design makes greater voltage, but a larger resistance weakens the output. The magnet should provide electromotive pressure in a series of inductors in garment additives. This helps the generator run well and gain energy. It can happen when clothes react rapidly and closely (Fig. 7).

This research examines the link between voltage impulses and their source to determine how many inductors can be placed in a magnet's path [11, 12]. By putting four inductors on a magnet trajectory and observing the relative rise in short-term strength, it was proven that three inductors can produce additive voltage impulses. Many garment prototypes were produced for real-life results; some are discussed below. Both watched someone walking at different rates. Because the best voltage and electricity depend on how rapidly the magnet moves and how close it is to the inductor, it should be placed between an overcoat's sleeve and shoulder (Fig. 8).

Figure 9 illustrates the inductive elements and the magnet. Electromagnetic human motion energy harvester parts: The inductive element is made up of three groups of spiral-shaped coils [12, 13]. Each coil contains five layers, with insulation in between. 2.5 cm broad, 50-turn coils [13]. A user set up the generator to view as they went. Normal voltage pulse form is governed by how the wearer's arms move. When the magnet near the coil completes one cycle, two pulses are created. Each arm's double-step movement cycle consists of forward and opposing actions. Pulses vary with sleeve motion. When the sleeve is dragged forward, the magnet's path is closest to the coils, increasing the pulse's voltage.

The magnet moves with the sleeve as it's pushed back. This reduces voltage. Herbal hands and garments disrupt impulses on both sides [13]. Because an electromagnetic harvester creates the most power while the wearer moves, it may not power a wearable digital device for long. Adding electricity harvesters that utilize human energy waste can boost the total supply.

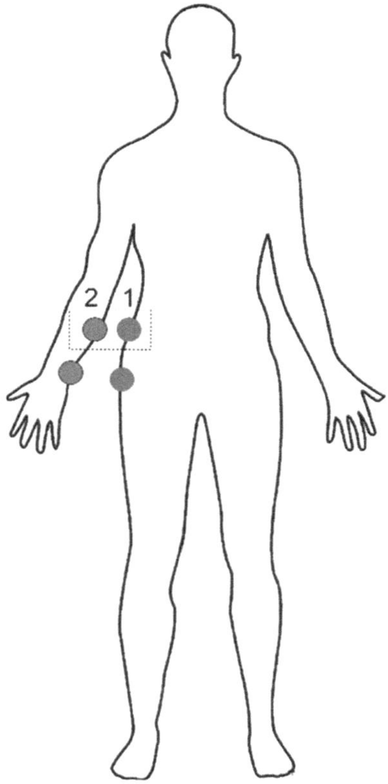

Fig. 8 Electromagnetic harvester placement in the human body

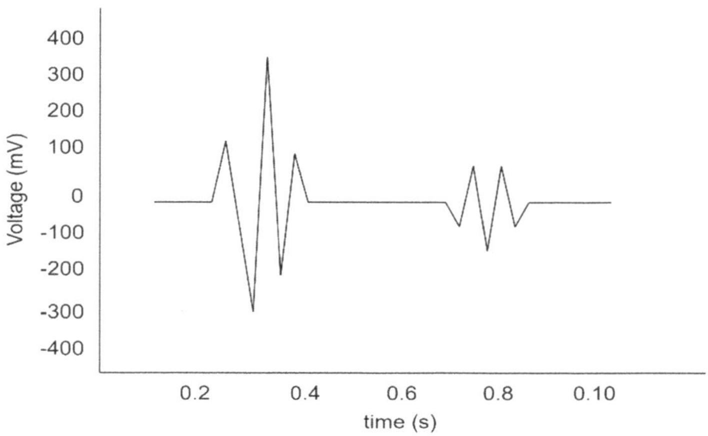

Fig. 9 Voltage pulse generated at walking speed

4 Biochemical Energy Harvesting

A biofuel cell converts electricity to fuel. Electricity is made by mixing anode and cathode oxidation and reduction. Catalysts accelerate chemical reactions. Biofuel cells can keep producing electricity if chemicals are introduced. Glucose is key. It's found in bodily fluids and can power biofuel cells. Anodically oxidized glucose releases electrons and hydrogen ions. Outside electrons can pass the barrier, but not the substance's hydrogen ions. Before cathode oxidation, price neutrality [14]. Depending on the catalyst, glucose biofuel cells are enzymatic, microbiological, or abiotic. Enzymatic biofuel cells catalyze reactions.

Laccase and glucose oxidase boost biofuel cell chemistry. Enzyme-powered biofuel cells produce 1.3 mW/cm^2. A biofuel-mobile-using rat got 109 W/cm^2. Enzyme-fueled cells don't persist long, which makes sense [14]. Short-term power. Live microorganisms power biofuel cells. Fuel cells can self-restart. Probably not. A microbial biofuel cell holds 1 mW/cm^2. Living organisms can't fuel biofuel cells. Noble metals or activated carbon are inorganic catalysts. Figure 10 shows glycol-biofuel cells. Living matter has biochemical power. Living beings can produce hydrogen with or without photonic light. The sun or something that resembles the sun can provide these photonic electrical reserves. This technique produces hydrogen without light. Dark fermentation occurs when mild supply diminishes. Since they don't need photon-friendly parts, these reactors are more reliable and less expensive. Dark fermentation is beneficial. By using herbal waste to produce hydrogen, waste can be treated, stabilized, and put off, reducing infection risk [15]. Another example is hydrogen-producing dark-fermented wastewater (H2). The anode bacteria determine the device's operating temperature. Temperature affects connections, activation, and solution flow. Temperature increases bacteria energy needs and membrane clarity. Electricity increase [14]. The natural count

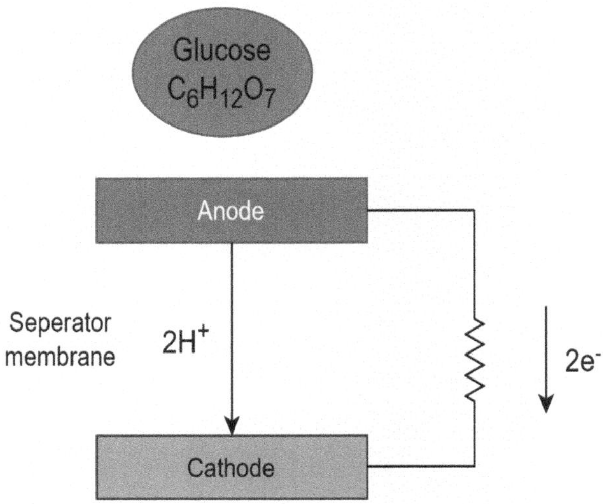

Fig. 10 Glucose biofuel cell

entering the molecule impacts the strength density because the reactor can only convert a specific amount of mass per unit area into energy and other products.

Interface circuitry must increase the voltage to power CMOS devices. Animals tested glucose-based biofuel cells. Three months after the implant, a rat experienced no pain or other difficulties. These devices need further research and information to be safe and effective [14, 15]. This biofuel cell isn't for sale, but it's used. This sparked fresh ideas. Microorganisms can create power from natural materials. They're working 70% of the time. Few electrons mean little power. This cell filters wastewater, unlike others. Cell electricity cleans water. It's half as energy-efficient.

5 Implementation of Energy Management in Biosensor

In the previous section, the best techniques to acquire energy from the body were picked based on their effectiveness and importance. Although high power densities are theoretically possible, lower electrical densities are more likely. Look at historical bankruptcies to see how powerful the most enticing claims were in reality. Under optimum conditions, biosensors may utilize tens of milliwatts (mW) or more [16]. More research is needed to understand biosensor circuits' electrical needs and behavior. This determines if self-powered biosensors are possible. Biosensor circuits and electrical harvesters must connect differently due to their peculiarities. Self-powered biosensors need a good electrical interface. It's crucial to know an interface circuit's basic features and specs. Biosensor circuitry and electricity-collecting gadget provide these.

6 Requirements of Biosensor

Information supplied and received quickly determines the demand for strength. Programs that analyze, process, and send a lot of data tend to use a lot of energy, whereas those that do so infrequently use less. When sending and receiving information, communication substructures use the greatest energy. Biosensors can selectively detect, send, and receive data since biosignals diffuse out slowly. Most biomedical goods [16] require few statistics. Biosensors can handle full duty cycling, which means they're mostly idle. During inactivity, little electricity is consumed (sleep phases). The working device rises swiftly and peaks when they're conversing. During a single cycle, energy needs to change from nW to mW [17].

Sleep affects how much energy is consumed on average. Consider how often an implanted biosensor measures glucose. The average amount of energy required by a sensor might be decreased by more than 5 times if it just had to calibrate and send data every 5 min instead of every minute [17]. Because a biosensor uses no power when off. As a result, $P_{\text{sleep}} = 0$ has the common electricity shown below:

$$P_c = D.P_{\text{active}} \tag{1}$$

If the responsibility cycle is shortened, common strength is used less (or, in similarly, extending the length of the sleep state). Some apps may not allow this since they have stringent running schedules [17, 18]. The biosensor needs a continuous voltage to work. Biosensors are complex devices with several RF, analog, and digital sub-blocks; therefore, they need multiple high-voltage energy sources. Implanted biosensors and other low-power applications benefit from several power sources [18]. Voltage can be altered on the building blocks. Only half of the voltage needed by digital circuits must be transmitted over RF. Virtual circuits feature high feedthrough currents and switch. They're also noisy. RF circuits and other analog circuits need a "clean" supply voltage. Having more assets gives you power, so that's wonderful.

7 Power Autonomy

Biosensors that are implanted might be able to get their own power from electricity harvesting, so they wouldn't need batteries [19]. Seems strange that a biosensor could be sending out data when the total strength is at its lowest. Because of these reasons, most energy-harvesting systems, especially those that are meant to be implanted, need to have a power garage tool built in (Table 1).

The thing that stores energy should also be able to work as a shock absorber. When there is a bigger need for power, like when it is sleeping, it should be able to give more power. If not, it should use less energy. Supercapacitors, new thin film batteries, and regular rechargeable batteries can all be used in power storage systems. Everyone knows that traditional batteries have a lot of problems. Thin film batteries are different from regular batteries because they are made of plastic, can be made in any shape or size, and can be built into IC packages. They have less electrical density than ordinary batteries. Supercapacitors have out-of-this-world capacitance. They work like batteries and capacitors. Supercapacitors store less electricity than batteries. They're more powerful per area. This makes supercapacitors useful for implanted biosensors, among other things. They can live a long time because they have so many ability cycles.

Table 1 Comparison of battery and supercapacitor parameters

Function	Lithium-ion	Supercapacitor
Cell voltage	2.3–2.75 V	3.6–3.7 V
Charge time	1–10 s	10–60 min
Specific energy	5 (Wh/kg)	100–200 (Wh/kg)
Cycle life	1 million or 30,000 h	500 and higher
Charge temperature	−40 to 65 °C	0 to 45 °C

8 Strategies for Harvesting Energy in the Human Body

In the past few years, a lot more people have become interested in wearable technology and how it can be used. Many studies on wearable technology have come up with a wide range of successes and failures. Wi-Fi sensor devices that can be attached to the body in different ways have become popular in the health, fashion, and sports industries. This is because companies are interested in making wearable consumer electronics, and many of these programs work together through smartphones.

8.1 Wearable Techniques of Energy Harvesting

Self-sustaining systems that obtain their energy from movement, the environment, or heated temperatures are interesting since they must act on the human body for lengthy periods of time [18]. EH technology can aid self-keeping in various ways. Collecting energy makes a self-running tool. Portable and wearable devices need clean, renewable, and sustainable energy to be comfortable and benefit the environment. There are several types of energy in the environment, including human and environmental energy. Figure 11 displays biomedical wearables approaches. Walking, running, and moving fingers can generate mechanical electricity. Human motion can happen in numerous ways and at low frequencies, making it harder to measure mechanical electricity.

Triboelectric and piezoelectric generators create mechanical electricity from human movement. Friction electrification simplifies triboelectric power collection [18, 19]. Imprinted prices don't change. Because the two charged objects are moving relative to one another, their capacities will differ. TENGs are the most popular

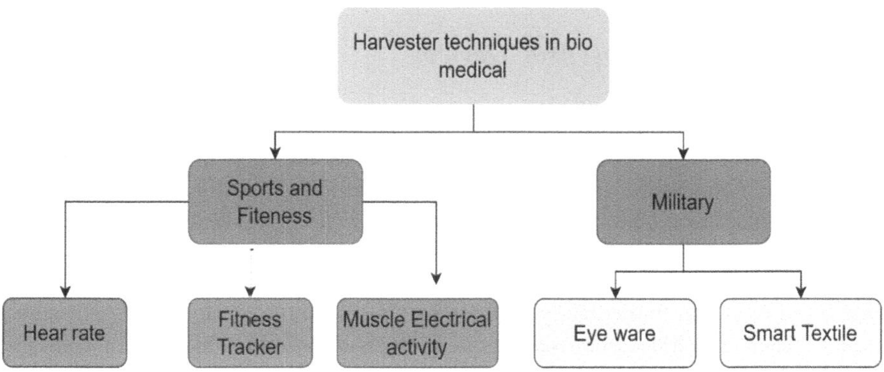

Fig. 11 Harvesting techniques involved in biomedical wearable devices

Table 2 Power is drawn from wearable component devices

Component	Characteristics	
	Ultralow power	Typical
Microcontroller	60 μA/MHz	225 μA/MHz
Sensors	100 μA	1–10 mA
Transceiver	6 mA	15 mA
Ultralow power communication	0–100 μA	1 mA

approach to picture human movement because they can be manufactured from a variety of materials, operate at a low frequency, and provide a lot of power.

Many wearables and portable, self-powered sensors for monitoring physical, chemical, and physiological data use energy-harvesting technologies. Self-powered sensing can be identified in several ways. Active sensing employs the electrical signal itself as the sensing signal and lets other factors alter the output. In tracking systems, active sensing monitors pressure, humidity, and temperature. A self-powered strain sensor created from graphene and zinc cable is also redox driven. Stretching graphene makes it more stress-resistant, lowering its redox side. This sensor can tell if the knee is actively moving.

In power-gathering devices, an electrode is covered to collect the electric field (20–80 MHz) on the skin's surface. By adjusting the transmitter-receiver distance, researchers mapped frame-wide power. At 2.44 GHz and 900 MHz, human bodies have an RF power switch that can adjust received power by 70 and 50 dB [20] (Table 2).

Unlike RF, the researchers were able to transmit 1.2 mW from one wrist to the other (a distance of 120 cm). Progress region insurance was increased by 1.1 W. Self-powered sensors can perform more now [20]. The self-powered electrical system must overcome challenges including long-term stability, multimode sensing, and electricity harvesting before being deployed on a big scale. Figure 12 depicts the wireless node structure.

8.2 *Implantable Biosensor Devices*

Biosensors can be used to measure and track people's behavior. When partially or totally implanted into a person's body, they are called "implantable biosensors." Implantable technology can lessen a person's pain and suffering. These implanted electronics may soon be used in biomedicine. They can give a clearer image of what happens inside the body over time, making it easier to track long-term illnesses, treatments, or procedures. The brain, heart, eyes, and blood contain them. Implantable biosensors can immediately indicate natural metabolites, electrically stimulate nerves, detect electrical impulses, repair biological systems, and distribute drugs. Blood pressure monitoring is vital for all organs [21]. Figure 13 shows implantable technologies in human body.

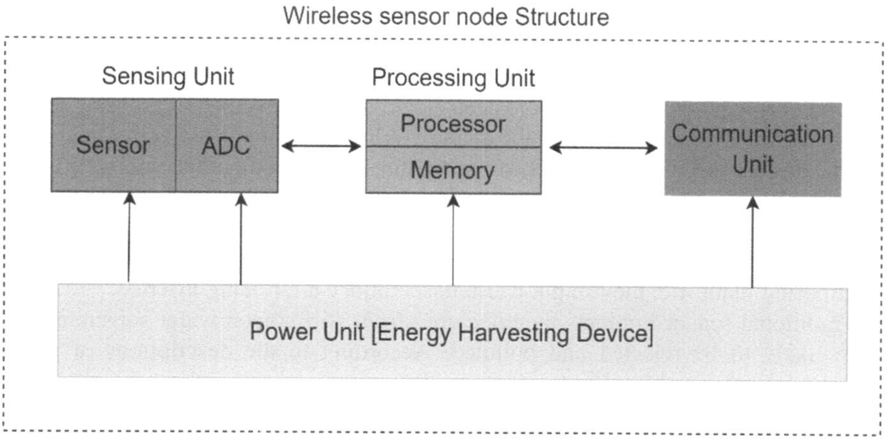

Fig. 12 Working on biomedical wearable device

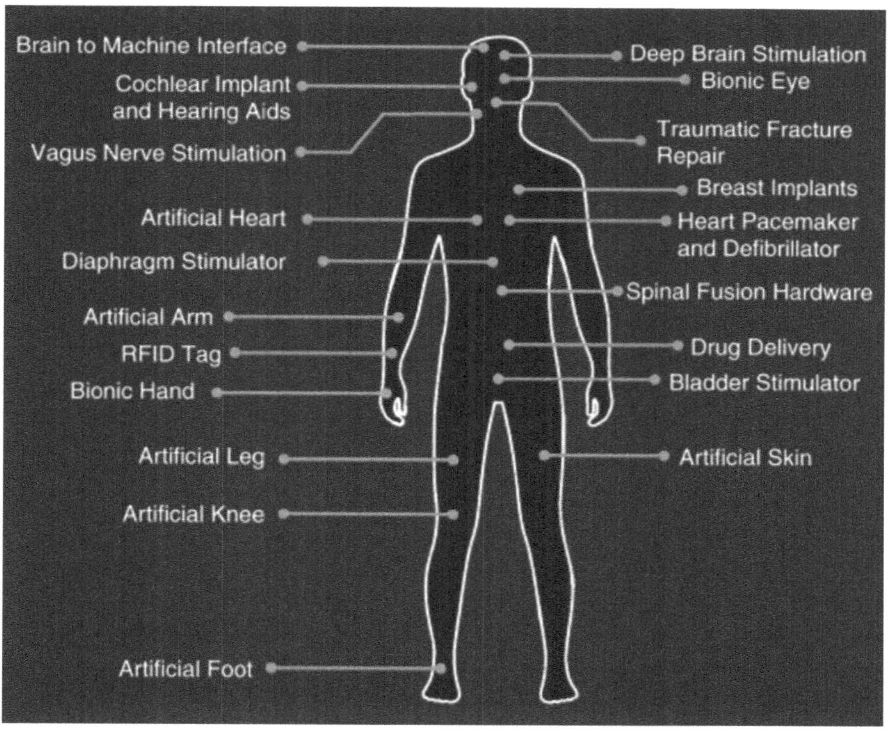

Fig. 13 Implantable technologies in human body

Pressure can be increased to stop or damage a bodily component. Coronary heart muscle disease can cause high blood pressure and infarction. Researchers are developing implanted blood pressure sensors to monitor and treat excessive blood pressure. Fully implanted biosensors need electrodes for detecting and sensing target analytes/critical signals, a circuit for displaying measurements and communicating results, and an electrical supply. Biosensors must be biocompatible and well tolerated by the body to avoid toxicity and long-term inflammation. Soft, curving human flesh is considerably different from rigid, flat semiconductor wafers [21]. Due to their sharp edges, stiffness, shape, and size, as well as the daily stress they inflict on neighboring plant life, they might cause tissue injury after being inserted.

Traditional sensor implants manufactured from stiff silicon wafer substrates are more likely to be rejected and polluted. According to the descriptions of these materials, fibrous tablets from all over the system, causing the sensor to fail in vivo. Biocompatible, flexible, bending substrates constructed of biopolymers should replace silicon wafers in medical applications to prevent the body-outside problem and fibrotic tissue encapsulation. Polymer substrates fill the area between soft, curving tissues and flat, stiff semiconductor chips. Never before has the implanted instrument needed to be so small [22].

Needle-inserted biosensors generate minimal tissue injury, inflammation, and immune system response. Miniaturization is possible via shrinking sensing electrodes, electric circuitry for power production and data exchange, and how they're assembled and packed. Using photolithography, dip-pen nanolithography, and micromachining, components can be coated and made smaller, leading to nanotechnology. When a biosensor is implanted, biofouling and FBR may occur. Due to tissue damage and inadequate biocompatibility, the device may be worthless [23–25]. Size, form, design, roughness, geometry, solubility, composition, interface fabric, sterilization, the timing of placement, packing, and degradation may also contribute to the body's unfavorable reaction.

Normal wound healing reactions, acute and continuous inflammation, granulomatous tissue development, and excessive fibrosis make up FBR (foreign body removal). Unspecific proteins from blood and tissue fluids grip or consume things during the start of a tissue-device interaction. Leukocytes, monocytes, and platelets may fight back and safeguard the body. The acute phase can last a few hours to 3 days. A frequently used implanted device creates continuous contamination. Macrophages, monocytes, and lymphocytes are active, and blood vessels and connective tissue are reorganized. More blood vessels help provide nutrients and repair wounds [26, 27]. Digital generation is predicted to get better and utilize less power, allowing for renewable energy systems. Always popular, harvesting techniques are always used. Post-implant monitoring aids in patient recovery and device safety. Remote monitoring enables in-depth data analysis, daily computerized transmissions, and long-term care while the patient is at home.

9 Conclusion and Future Technology

Portable gadgets use biomechanical energy harvesting. We've utilized biomechanical models to predict how much electricity each human activity can produce and discussed its pros and cons. TFET-based biosensors have real-time performance improvements over standard FET-based biosensors. From its earliest beginnings to its most recent improvements, TFET is evaluated as a biosensor utilizing multiple methodologies [28–30]. The evaluation shows that identifying proteins with changing dielectric constants could be a future study field. Less doping and charge plasma development may have eliminated doping difficulties and facilitated production. Horizontal and vertical tunneling may also affect burrowing. There are still more responsiveness boundaries to identify, which might be used to develop TFET-based biosensors and determine their sensitivity to limit voltage and subthreshold swing variations. Implantable scientific equipment uses energy harvesting to replace batteries. It's better than batteries. These improvements could lead to a breakthrough in personal scientific electronics and implant development.

References

1. Rajavi Y, Taghivand M, Aggarwal K, Ma A, Poon ASY (2016) An energy harvested ultra-low power transceiver for Internet of Medical Things. In: Proceedings of IEEE European solid-state circuits conference, pp 133–136
2. Bhushan B (2011) MEMS/NEMS and BioMEMS/NEMS: materials, devices, and biomimetics. In: Nanotribology and nanomechanics: nanotribology, biomimetics and industrial applications. Springer, Berlin, pp 833–945
3. Billinghurst M, Starner T (1999) Wearable devices: new ways to manage information. Computer 32(1):57–64
4. Khaligh PZ, Zheng C (2010) Kinetic energy harvesting using piezoelectric and Electromagnetic technologies—state of the art. IEEE Trans Ind Electron 57(3):850–860
5. Shenck NS, Paradiso JA (2001) Energy scavenging with shoe-mounted piezoelectrics. IEEE Micro 21(3):30–42
6. Dimitrov DV (2016) Medical internet of things and big data in healthcare. Healthc Inform Res 22(3):156–163
7. Wise K, Anderson D, Hetke J, Kipke D, Najafi K (2004) Wireless implantable microsystems: high-density electronic interfaces to the nervous system. Proc IEEE 92(1):76–97
8. Magno M, Boyle D (2017) Wearable energy harvesting: from body to battery. In: 2017 12th International conference on design & technology of integrated systems in nanoscale era (DTIS), pp 1–6
9. Shaikh FK, Zeadally S (2016) Energy harvesting in wireless sensor networks: a comprehensive review. Renew Sust Energ Rev 55:1041–1054
10. Feenstraa J, Granstroma J, Sodano H (2008) Energy harvesting through a backpack employing a mechanically amplified piezoelectric stack. Mech Syst Signal Process 22(3):721–734
11. Mitcheson PD, Yeatman EM, Kondala Rao G, Holmes AS, Green TC (2008) Energy harvesting from human and machine motion for wireless electronic devices. Proc IEEE 96(9):1457–1486
12. Goto K, Nakagawa T, Nakamura O, Kawata S (2001) An implantable power supply with an optically rechargeable lithium battery. IEEE Trans Biomed Eng 48(7):830–833

13. Jow U, Ghovanloo M (2007) Design and optimization of printed spiral coils for efficient transcutaneous inductive power transmission. IEEE Trans Biomed Circuits Syst 1(3):193–202
14. Liu W et al (2005) Implantable biomimetic microelectronic systems design. IEEE Eng Med Biol Mag 24(5):66–74
15. Sodagar AM, Wise KD, Najafi K (2009) A wireless implantable microsystem for multichannel neural recording. IEEE Trans Microwave Theory Tech 57(10):2565–2573
16. Torfs T, Leonov V, Yazicioglu R, Merken P, Hoof CV, Vullers R, Gyselinckx, B (2008) Wearable autonomous wireless electroencephalography system fully powered by human body heat. IEEE Sens:1269–1272. https://doi.org/10.1109/ICSENS.2008.4716675
17. Saravanan M, Ajayan J, Maheswar R, Eswaran P (2019) An overview of architecture & applications of IoT based health care system, book: smart healthcare systems: applications, services, and challenges. Wiley Telecom. https://doi.org/10.1002/9781119816829.ch5
18. Kim S, Vyas R, Bito J, Niotaki K, Collado A, Georgiadis A, Tentzeris MM (2014) Ambient RF energy-harvesting technologies for self-sustainable standalone wireless sensor platforms. Proc IEEE 102(11):1649–1666
19. Calhoun BH, Khanna S, Zhang Y, Ryan J, Otis B (2010) System design principles combining sub-threshold circuit and architectures with energy scavenging mechanisms. In: Proceedings of IEEE International Symposium on Circuits and Systems (ISCAS), pp 269–272
20. Ahmadi MM, Jullien GA (2009) A wireless-implantable microsystem for continuous blood glucose monitoring. IEEE Trans Biomed Circuits Syst 3(3):169–180
21. Clausen I, Glott T (2014) Development of clinically relevant implantable pressure sensors: perspectives and challenges. Sensors 14:17686–17702
22. Anuradha K, Saravanan M, Sathya K, Prabhu N, Karthik M (2016) Design and development of a portable ECG acquisition system using cadence. Adv Nat Appl Sci 10(3):83–89
23. Vaddiraju S, Tomazos I, Burgess DJ, Jain FC, Papadimitrakopoulos F (2010) Emerging synergy between nanotechnology and implantable biosensors: a review. Biosens Bioelectron 25:1553–1565
24. Wang Y, Vaddiraju S, Gu B, Papadimitrakopoulos F, Burgess DJ (2015) Foreign body reaction to implantable biosensors: effects of tissue trauma and implant size. J Diabetes Sci Technol 9: 966–977
25. Saravanan M, Parthasarathy E (2021) A review of III-V tunnel field effect transistors for future ultra-low power digital/analog applications. Microelectron J 114:105102
26. Saravanan M, Parthasarathy E (2021) Investigation of RF/analog performance of InAs/InGaAs channel based nanowire TFETS. In: 2021 International Conference on Communication, Control and Information Sciences (ICCISc), pp 1–4
27. Saravanan M, Parthasarathy E (2021) Investigation of RF/Analog performance of Lg=16nm Planner In0.80Ga0.20As TFET. In: 2021 Fourth International Conference on Electrical, Computer and Communication Technologies (ICECCT), pp 1–4
28. Nagendra Reddy N, Panda DK (2021) A comprehensive review on Tunnel Field-Effect Transistor (TFET) based biosensors: recent advances and future prospects on device structure and sensitivity. SILICON 13:3085–3100
29. Avula M et al (2013) Modulation of the foreign body response to implanted sensor models through device-based delivery of the tyrosine kinase inhibitor, masitinib. Biomaterials 34: 9737–9746
30. Mohanbabu A, Saravanan M, Ajayan J, Basharan S (2021) Design and development of AlGaN/ GaN HEMT for bio sensing applications. In: Electronic devices, circuits and systems for biomedical applications. Elsevier. ISBN: 9780323851725

A Dual-Band RF Energy Harvesting System in Sub-6 GHz for Low-Power Electronic Appliances

Srinivasu Garikipati, Gayatri Tangirala, Pavan Mehta, Gaurav Varshney, Manikya Krishna Chaitanya Durbhakula, and Virendra Kumar Sharma

1 Introduction

Huge EM energy from the base stations and their wireless services are unutilized. This energy creates the need of RF energy harvesting (RFEH) system. In recent days, RFEH techniques play a major role in powering low-power electronic devices, biomedical implanted devices, mobiles, IoT devices, electric vehicles, and self-sustainable modules. The battery maintenance, cost, life cycle, and its replacement are the challenges for low-powered appliances [1–4]. In the early twentieth century, Tesla's experiments proposed the concept of EM energy transformation [5, 6]. This leads to an introduction of rectenna [7] that consists of the antenna for unused RF power sensing and a RF rectifier to convert this RF to DC power [8–10]. The performance of the rectifier is decided by a factor called power conversion efficiency (PCE). Various rectifier topologies are implemented to achieve the higher PCEs. There are mainly four types of rectifier topologies. They are single series [11], single shunt [12], voltage doubler [13], and bridge type [14]. Based on the operating frequency, the rectifiers are categorized into three types [15]. They are single band

S. Garikipati (✉) · G. Tangirala · V. K. Sharma
Department of Electronics and Communications Engineering, Bhagwant University, Ajmer, Rajasthan, India

P. Mehta
School of EEE, VIT Bhopal University, Bhopal, Madhya Pradesh, India

G. Varshney
Department of Electronics and Communications Engineering, NIT Patna, Patna, Bihar, India
e-mail: gaurav.ec@nitp.ac.in

M. K. C. Durbhakula
Department of Electronics and Communications Engineering, Vasavi College of Engineering, Hyderabad, Telangana, India

A. Nella et al. (eds.), *Energy Harvesting Trends for Low Power Compact Electronic Devices*, EAI/Springer Innovations in Communication and Computing,
https://doi.org/10.1007/978-3-031-35965-1_5

[16], multiband [17–20], and broadband [14]. As the ambient RF power density collected by RF harvester is relatively small, the designing of a RFEH system with a high RF to DC PCE is in great demand [21, 22]. In case of MIMO antenna, the challenges are to improve the channel capacity, acquiring isolation between the radiating elements < -15 dB, low channel capacity loss (CCL), high diversity gain (DG), unity mean effective gain (MEG) ratio, low envelope correlation coefficient (ECC), and stable total active reflection coefficient (TARC) [23]. Huge literature is available on ultra-wideband MIMO antennas. To improve the isolation between the radiators, various techniques were introduced. They are neutralization lines (NL) [24], decoupling networks (DN) [25, 26], electronic bandgap (EBG), metamaterial (MTM) [27, 28], defected ground structures (DGS) [29, 30], complementary split ring resonator (CSRR) [31], slots [32, 33], insertion of stubs [34], frequency reconfigurable [35, 36], inter-element spacing [37, 38], etc.

The major contribution of this chapter is a dual-band rectifier with enhanced PCE for proposed diamond slotted MIMO antenna at 1.8GHz and 3.6GHz, i.e., 5G NR n3 and n78 bands, respectively, for ambient RF energy harvesting. These frequencies offer higher values of RF power density that are from 36 to 84 nW/cm^2 [3]. The organization of the chapter consists of the following: Section 2 provides the evolution of diamond slotted MIMO antenna, Section 3 outlines the antenna results, and Sect. 4 reports the RF rectifier topology. Section 5 discusses proposed RF harvester results and Sect. 6 presents the conclusions of the work.

2 Design Evolution of Presented Dual-Band RFEH System

Firstly, a single-unit diamond ultra-wideband antenna was implemented. Then a four-element MIMO is developed by an orthogonal arrangement of radiators with 3 mm inter-element spacing. Later this MIMO module is implemented with novel DN for improvement of isolation among the radiators. Thereafter a DBIMN is implemented to minimize the complexity of rectifier and plays a vital role for enhancement of RF to DC PCE. Then a RF to DC signal converter and a DC to DC booster circuit are implemented with enhanced RF to DC PCE. The antenna simulations are performed on Ansys HFSS software and rectifier simulations are on ADS tool.

2.1 Single-Unit Diamond Slotted Antenna

Figure 1a shows the single unit of a high-gain diamond slotted antenna [39] design with optimized dimensions and working in the band 1.71–12GHz. The module is fabricated on the FR4 substrate of volume 44 × 34 × 1.6 mm^3 and it is shown in Fig. 1b. This module achieves a VSWR $\leq$1.9, $S_{11} \leq -10$ dB from 1.70 to 12.82 GHz shown in Fig. 1c, greatest peak gain as 7.25dBi, and maximum radiation efficiency as 97%.

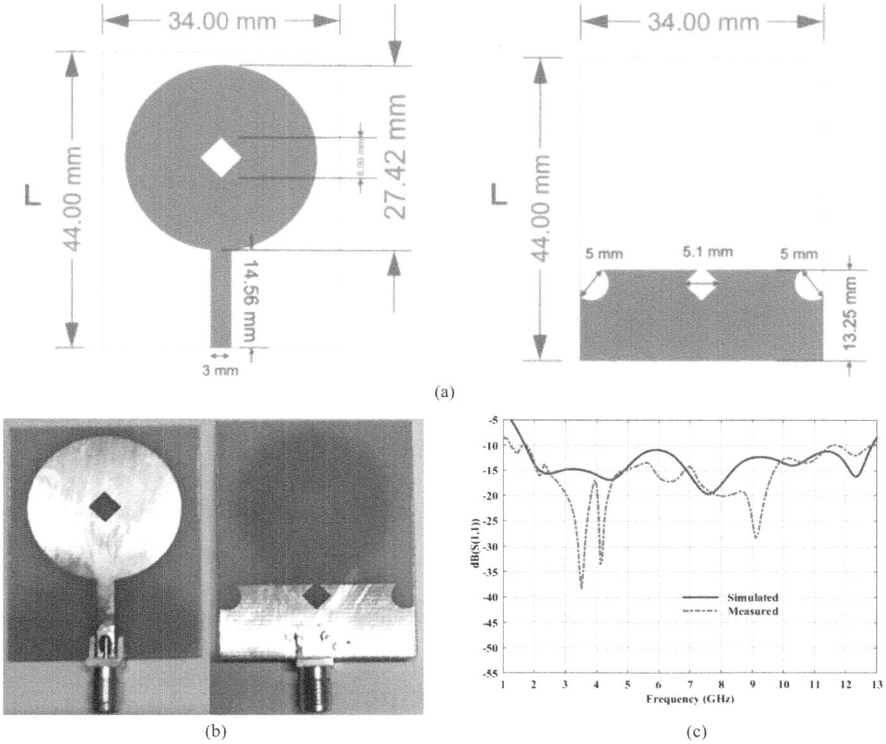

Fig. 1 Diamond slotted ultra-wideband unit. (**a**) Top and bottom views. (**b**) Printed module. (**c**) S_{11} plot

2.2 Development of Four-Port Ultra-Wideband MIMO Antenna

The four-port MIMO design is shown in Fig. 2 and is achieved by the orthogonal arrangement of radiators. Usually the spacing among radiators is $\lambda/4 \leq d \leq \lambda/2$ where λ is the lower resonating frequency to achieve optimum results. This MIMO is developed for 3 mm inter-element spacing with volume $81 \times 81 \times 1.6$ mm^3. The performance characteristics are impedance bandwidth is 1.4–13 GHz, peak gain is 6.1 dBi, isolation coefficients are S21 ≤ -22 dB, S31 ≤ -16.5 dB, and maximum efficiency of MIMO is obtained as 96%.

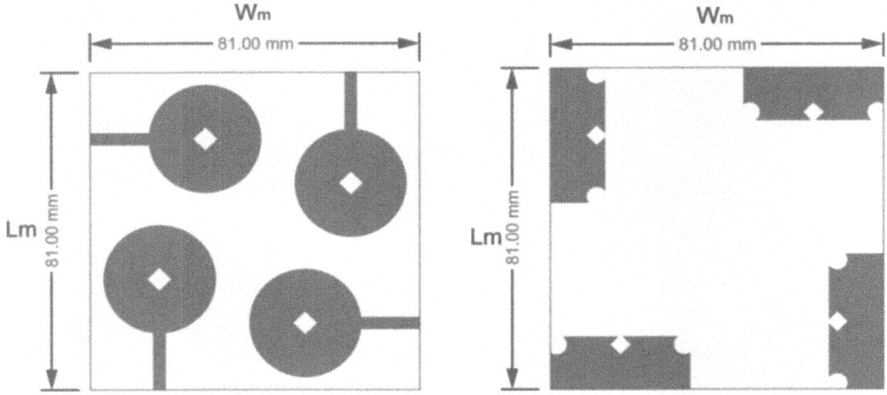

Fig. 2 Top and bottom views of four-port MIMO

2.3 Development of Four-Port Ultra-Wideband MIMO Antenna with DN

The four-port MIMO module is then verified with a novel DN and is represented in Fig. 3a for the purpose of enhancement of isolation among the radiators, and its fabricated prototype is represented in Fig. 3b. The measurement picture is represented in Fig. 3c. The performance characteristics are impedance bandwidth is 1.38–12.88 GHz, peak gain is 7.3 dBi, isolation coefficients are S21 ≤ -22.23 dB, S31 ≤ -17.41 dB, and maximum efficiency of MIMO is obtained as 97%.

3 Results and Discussion of Printed Four-Port MIMO with DN

The design simulations are performed with the Ansys HFSS software. The testing and measurement is done using VNA and anechoic chamber. The observations are documented in subsequent sections.

3.1 Return Loss and Mutual Coupling Coefficients

The printed MIMO scattering parameters comparison plots of S_{11}, S_{21}, and S_{31} are depicted in Fig. 4. It is noticed that $S_{11} \leq -10$ dB in the range 1.38–12.88 GHz, $S_{21} \leq -22.23$ dB, and $S_{31} \leq -17.41$ dB. The plots revealed that the measured and simulated results are almost similar.

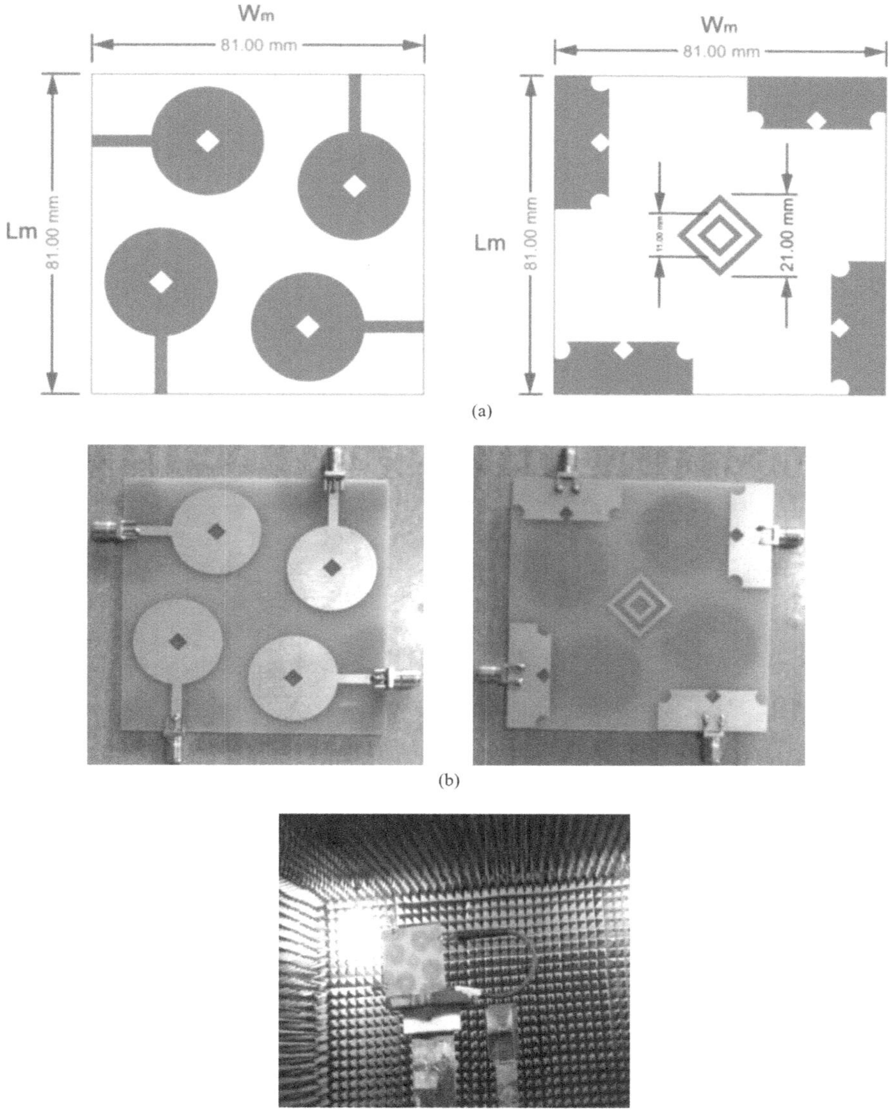

Fig. 3 Four-port MIMO with DN. (**a**) Top and bottom views with dimensions. (**b**) Printed module. (**c**) Measurement picture

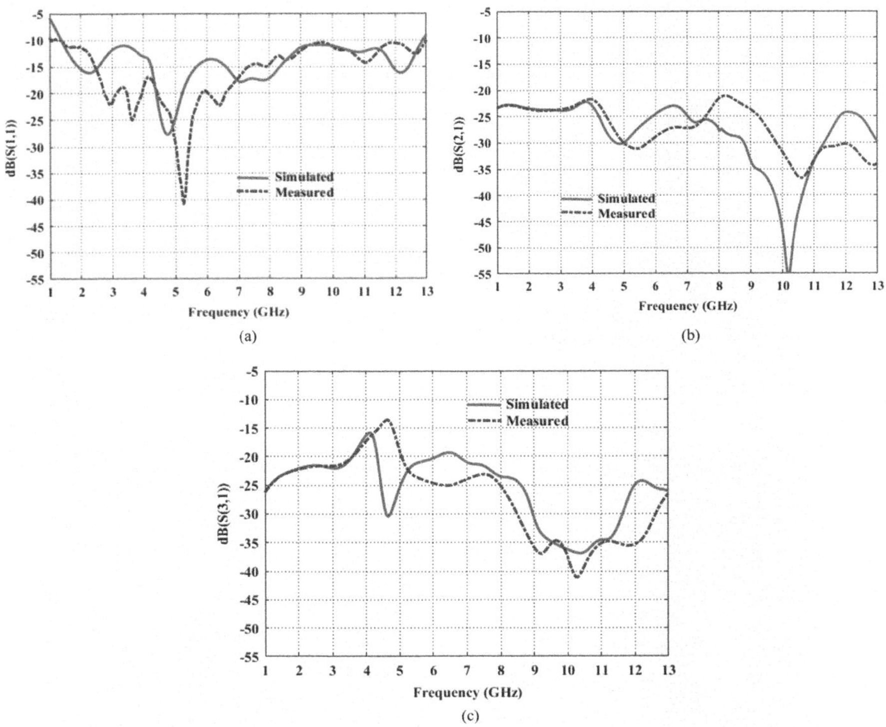

Fig. 4 Printed MIMO and their comparison graphs: (**a**) S_{11}, (**b**) S_{21}, (**c**) S_{31}

3.2 Radiation Characteristics

The 2D radiation patterns and its comparison graphs of the proposed MIMO for $\phi = 0°$ and $\phi = 90°$ at 4.6 GHz and 7.6 GHz are depicted in Fig. 5, respectively. The plots reveal that the measured results are almost similar to simulated results.

3.3 Gain and Efficiency Characteristics

Figure 6 depicts the peak gain and efficiency graphs of the printed MIMO by exciting port 1, while terminating other three ports with 50 Ω impedance. The highest gain reported is 7.3 dBi at 11.6 GHz. The highest efficiency is 97% at 1.9 GHz.

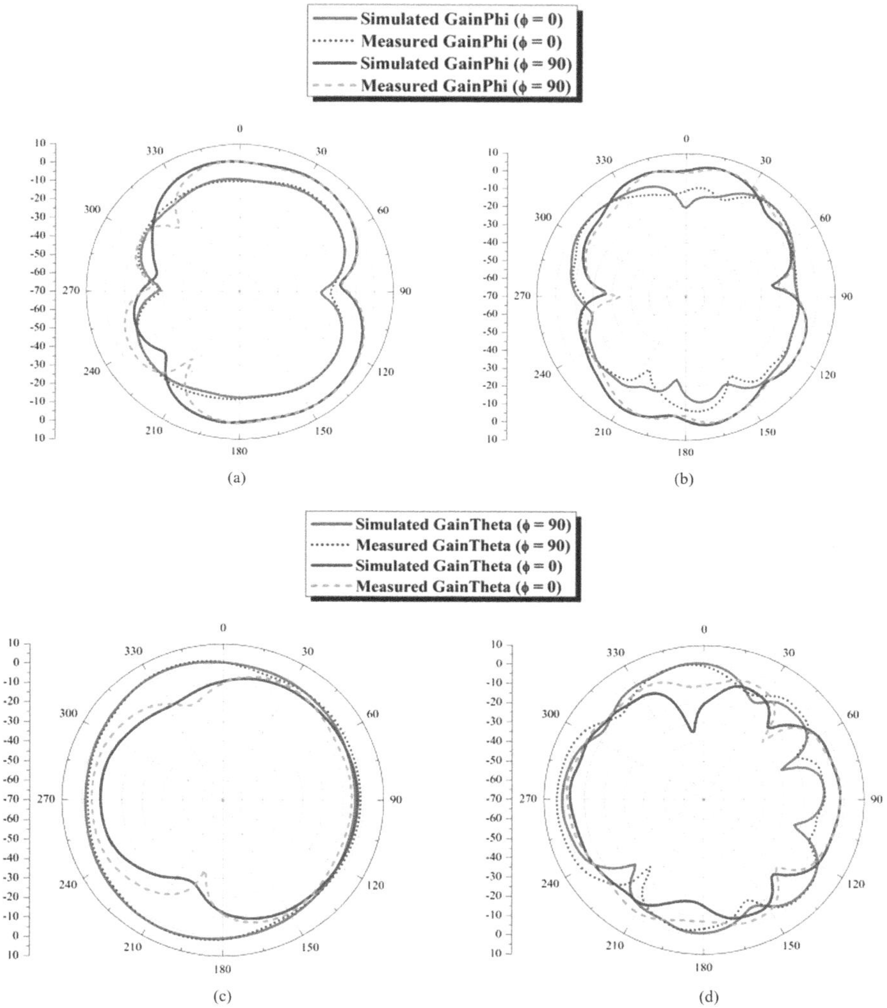

Fig. 5 Printed MIMO antenna 2D radiation performance for GainPhi at (**a**) 4.6 GHz and (**b**) 7.6 GHz; GainTheta at (**c**) 4.6 GHz and (**d**) 7.6 GHz

3.4 Diversity Analysis of Printed MIMO

ECC characterizes the correlation between the radiating elements. It is expressed by formula (1). Ideally it should be <0.5. Here, N is the number of radiators. DG $\approx$ 10 dB and it is given by formula (2). The proposed antenna shows ECC $\leq$ 0.0002, DG $\leq$ 10 dB.

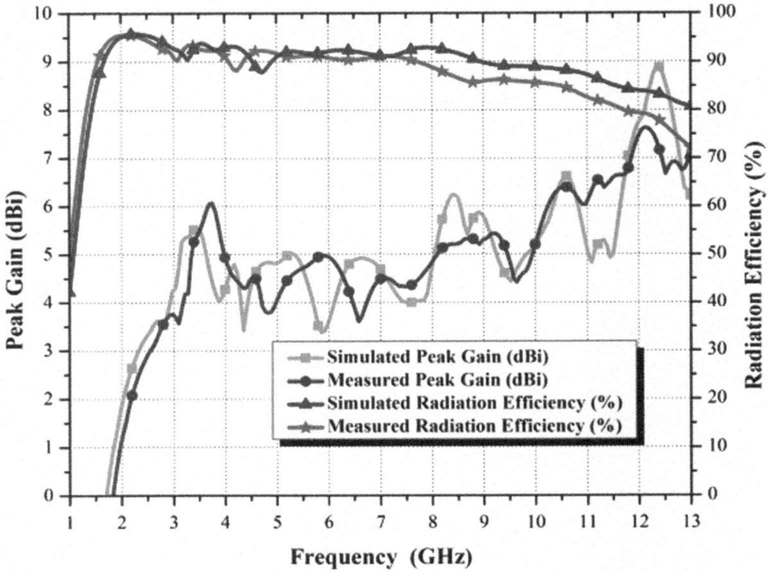

Fig. 6 Gain and efficiency plots of printed MIMO

$$\mathrm{ECC} = \rho_{ij} = \frac{\left|\sum_{n=1}^{N} S_{in}^{*} S_{nj}\right|^2}{\left[1 - \sum_{n=1}^{N} |S_{ni}|^2\right]\left[1 - \sum_{n=1}^{N} |S_{nj}|^2\right]} \quad (1)$$

$$\mathrm{DG} = 10\sqrt{1 - \mathrm{ECC}^2} \quad (2)$$

TARC defines the stability of MIMO system when the phases (ϕ) of input signal changed between 0° and 180° while first port is kept unchanged. It is expressed by formula (3). It is observed ≤ -10 dB for proposed MMO antenna.

$$\mathrm{TARC} = \frac{\sqrt{\sum_{i=1}^{4} \left|S_{i1} + \sum_{m=2}^{4} S_{im} e^{j\phi_{m-1}}\right|^2}}{\sqrt{4}} \quad (3)$$

The increment of radiators increases the capacity of the channel in MIMO while keeping transmission power levels or bandwidth constant. But there is interference between radiators that in turn causes CCL. The tolerable limit of CCL is 0.4 bps/Hz. MEG indicates the average received power from individual antenna units with reference to an isotropic antenna. It is expressed by formula (4). The CCL is $\leq$0.4 bps/Hz and MEG ratio $\approx$ 1 for the proposed MMO antenna.

$$\mathrm{MEG}_i = 0.5\left[1 - \sum_{j=1}^{\mathrm{N}} \left|S_{ij}\right|^2\right] \tag{4}$$

From Table 1 it is observed that the printed MIMO design is a good module for sensing unused RF energy in GSM1800, Wi-Fi, ISM, sub-6 GHz, and X-bands.

4 Rectifier Design

RFEH is the process of extracting power from the ambient RF sources and then converts to DC power. In this chapter, a rectifier with dual operating frequencies is designed with good impedance matching network for sub-6 GHz 5G NR n3/n78 bands. Figure 7a shows the topology of the presented DBIMN. Typically the RF input source has impedance of 50 Ω. Figure 7b shows a dual-band RF rectifier having a Schottky diode, a capacitor to filter DC, and a resistive load. The matching network DBIMN reduces the antenna losses for input RF power. The diode plays a major role in transforming the unused ambient RF energy to a useful DC power. The capacitor filter rejects the unwanted harmonics from the output of RF rectifier and stores the converted DC power. Generally load is complex quantity and depends on frequency.

The electrical schematic circuit of the presented dual-band AC to DC converter comprises of a DBIMN, a rectifier with two Schottky diodes HSMS-2850 and two capacitors, a capacitor filter, and a resistive load. The HSMS-2850 is a RF Schottky diode having 2 V peak reverse voltage, 0.15 V forward voltage, and 0.3 pF junction capacitance. These two diodes are placed in the voltage doubler form. The capacitor filter rejects the harmonics at the proposed frequencies 1.8 GHz and 3.6 GHz. After that a suitable resistive load is chosen to achieve higher PCEs. For this purpose a source-pull simulation is used to estimate the proper load value. The source-pull simulations have a single-tone AC input source with a DC block. Generally the dual-band design operates at two different frequencies; it needs an optimum resistance. Hence, the load $R_{\mathrm{L}} = 5$ K Ω is chosen as it provides good PCE for the input power $\mathrm{P_{in}} = 0$ dBm. The antenna and rectifier are matched to 50 Ω impedance. A DBIMN is achieved by lumped parameter-based LC matching network and it is used to obtain the dual-band performance. The input impedance of the rectifier changes with the frequency, input power, and load resistance due to the nonlinear behavior of the diode.

Table 1 Performance comparison of printed MIMO with recent work

Refs.	Volume of printed MIMO (mm^3)	Bandwidth (GHz)	Substrate	ECC	Isolation (dB)	Gain(dBi)	Radiation efficiency
[26]	85 × 85 × 1.27	3.69–13.89	Duroid	<0.01	<−21.87	4.8	NA
[33]	80 × 80 × 1.6	3–14	Duroid 5880	<0.03	<−20	4.8	>83%
[37]	80 × 80 × 0.764	2.6–11	FR4	<0.01	<−17.4	3.38	>85.7%
[40]	110 × 110 × 1.57	1.7–7.2	CER-10	<0.0025	<−20	5	NA
This work	**81 × 81 × 1.6**	**1.38–12.88**	**FR4**	**<0.0002**	**<−22.23**	**7.3**	**>77%**

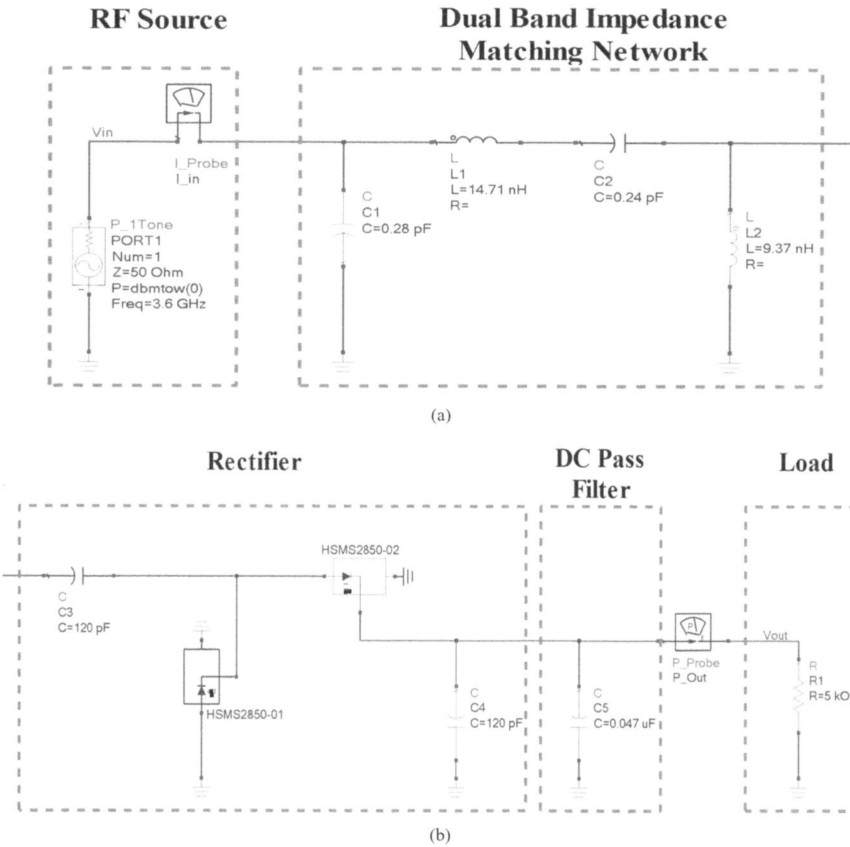

Fig. 7 The proposed electrical schematic of circuit. (**a**) DBIMN. (**b**) The dual-band AC to DC converter

5 Rectifier Results and Discussion

The rectifier circuit is simulated using ADS tool, and the performance is verified using co-simulation of harmonic balance and momentum at input power levels from -5 dBm to 5 dBm. The impedance matching network design process is achieved by lumped parameter-based LC matching network. Figure 8a depicts the simulated return loss, and 8b shows input impedance of the designed AC to DC converter with real and imaginary components of Z_{in} at various input power levels of rectifier with and without an impedance matching circuit. The return loss drops sharply at the matching frequency and input impedance of rectifier is matched to 50 Ω. Figure 8c shows reflection coefficient ≤ -10 dB at operating frequencies 1.8 GHz and 3.6 GHz.

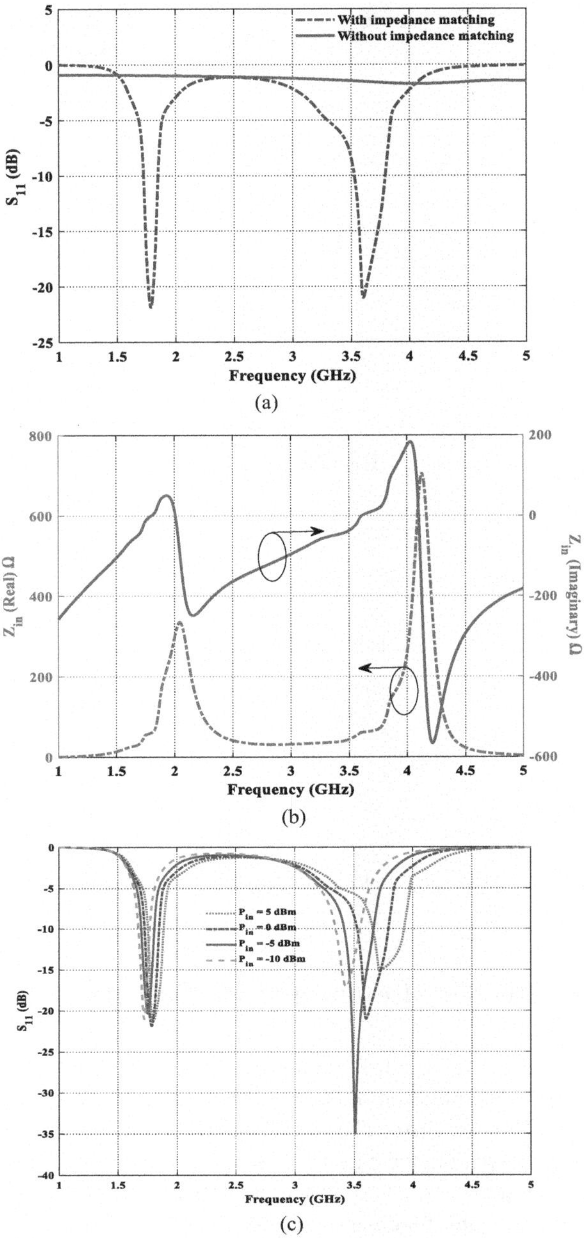

Fig. 8 Rectifier characteristics. (**a**) Reflection coefficient. (**b**) Input impedance at $P_{in} = 0$ dBm and $R_L = 5$ kΩ. (**c**) Reflection coefficient of AC to DC converter at various input powers and $R_L = 5$ kΩ

It is noticed from Fig. 8c $S_{11} \leq -21$ dB and hence there exists a good matching between antenna and rectifier. So, good conversion efficiencies and peak DC output voltages are achieved at proposed frequencies. From Fig. 9 it is observed that the rectifier attained the highest DC output voltages of 1.85 V and 1.6 V and enhanced RF to DC PCEs of 76.0% and 57.2% at 1.8 GHz and 3.6 GHz for $P_{in} = 0$ dBm with load 5 kΩ, respectively.

Figure 10 displays the simulated DC output voltage and PCE with respect to the input powers at various operating frequencies. The RF to DC PCE, η_{RF-DC}, is the ratio of the DC power transmitted on output load (R_L) to the collected RF input power. It is given by formula (5). It is observed from the graphs that the PCEs are more than 50% with the input powers from −7 to 7 dBm at 1.8 GHz. It is greater than 50% with the input powers from −4 to 6 dBm at 3.6 GHz.

$$\eta_{RF-DC} = \frac{P_{DC}}{P_{RF}} \times 100\% \tag{5}$$

Figure 11 provides the output DC voltage and PCE of the proposed dual-band AC to DC converter circuit at various input power levels. The P_{in} ranges from −10 to 10 dBm at proposed frequencies to evaluate the AC to DC PCE and output DC voltage at load $R_L = 5$ kΩ. Table 2 represents the comparison of the presented design with related work.

6 Conclusion

This chapter presents an RFEH system that comprises of diamond slotted MIMO antenna, a DBIMN, RF to DC signal converter, and DC to DC booster circuit for harvesting ambient RF energy for 1.8 GHz and 3.6 GHz, i.e., 5G NR n3 and n78 bands, respectively. The MIMO module is printed on an FR4 substrate of volume $81 \times 81 \times 1.6$ mm^3 to capture unused RF energy from sub-6 GHz bands. The antenna characteristics are $S_{11} \leq$ -10 dB in the range 1.38–12.88GHz, $S_{21} \leq -22.23$ dB, and $S_{31} \leq -17.41$ dB, ECC ≤ 0.0002, TARC ≤ -10 dB, DG ≤ 10, CCL < 0.4 bps/Hz, and MEG ratio ≈ 1. The highest peak gain is 7.3dBi and efficiency is 97%. The DBIMN is achieved by lumped parameter-based LC matching network. For validation, the proposed rectifier design is optimized for different power levels (−5 dBm to 5 dBm) using harmonic balance to operate at the frequencies 1.8 GHz and 3.6 GHz. The voltage doubler rectifies the captured RF energy using a pair of Schottky diodes SMS-7630 with capacitor and DC pass filter for harmonic rejection of RF signal. The measured PCEs are 72.8% and 62% at 5 dBm high input power and 56.2% and 44.1% at −5 dBm low input power for operating frequencies at 5 kΩ load. The presented design has high PCE and RF to DC output voltage is from 0.9 to 3 V. These values are sufficient for many low-power electronic appliances.

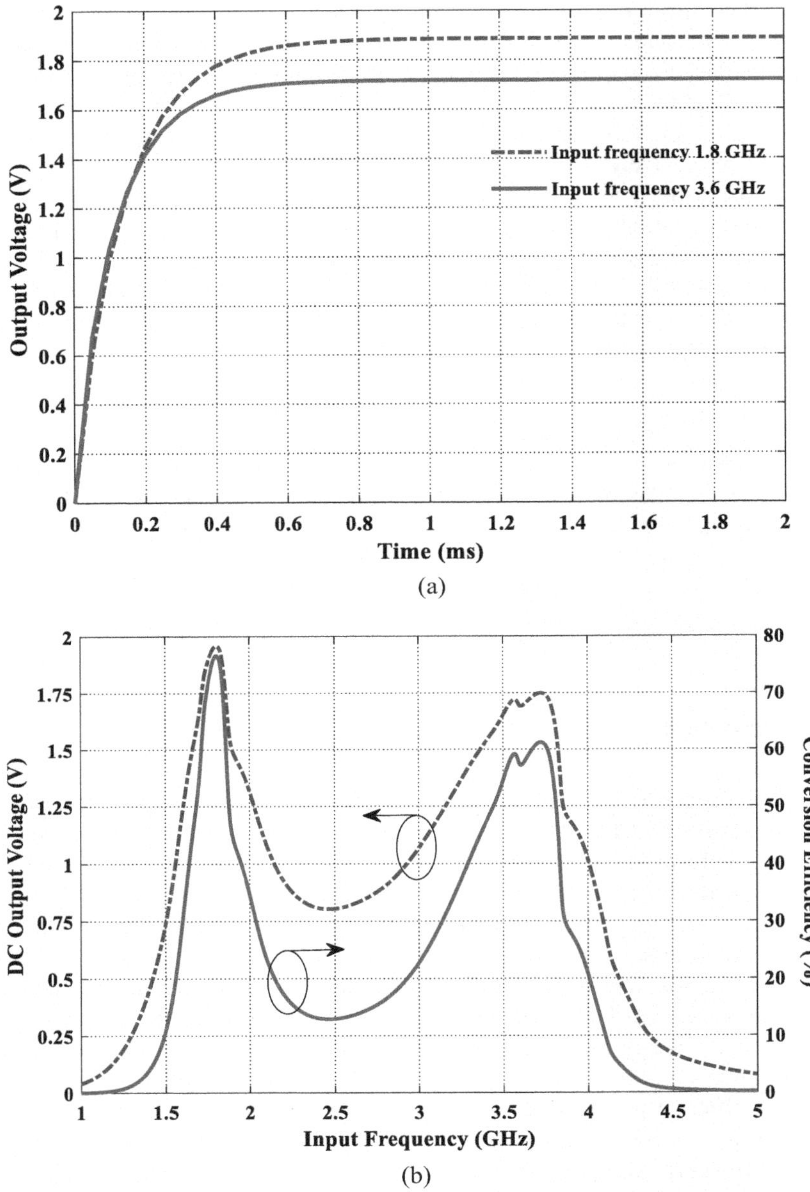

Fig. 9 (**a**) Output voltage at $P_{in} = 0$. (**b**) PCE and DC output voltage plots at $P_{in} = 0$ dBm and $R_L = 5$ kΩ

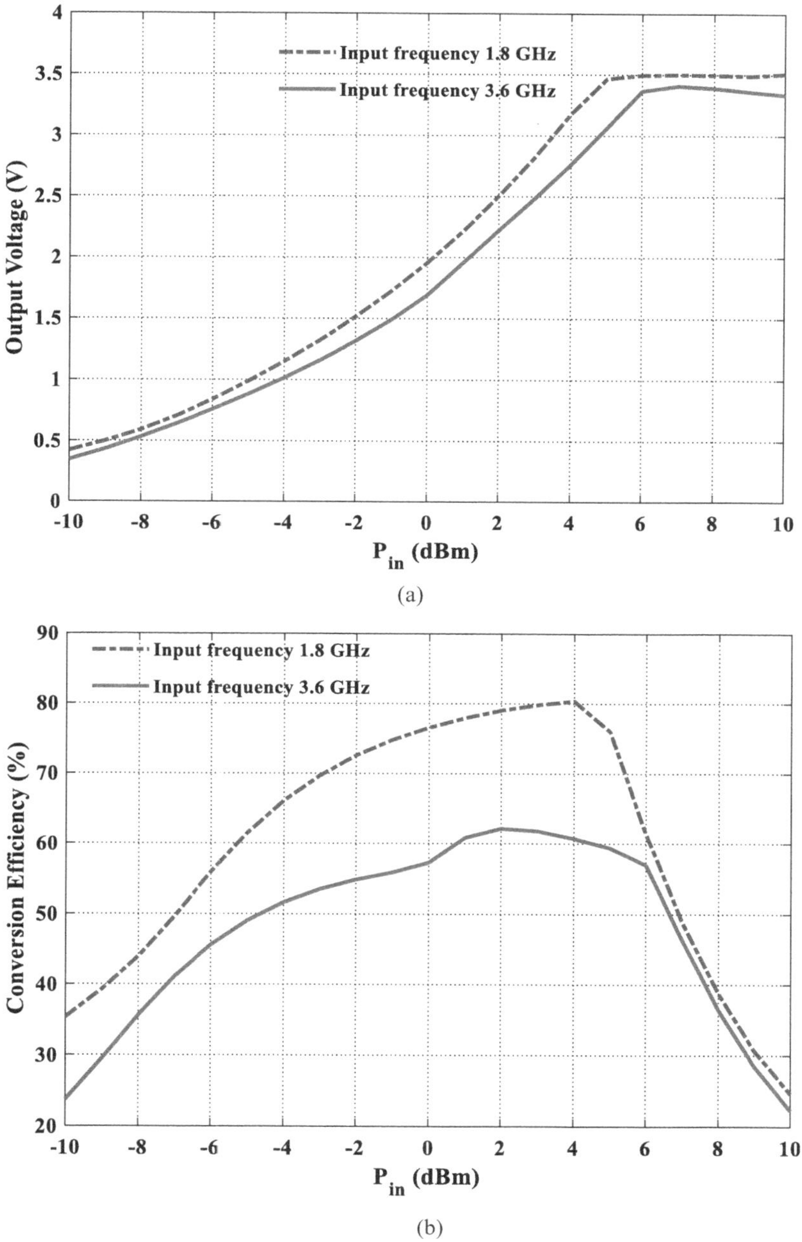

Fig. 10 (**a**) Output voltage. (**b**) PCE at $R_L = 5\ k\Omega$

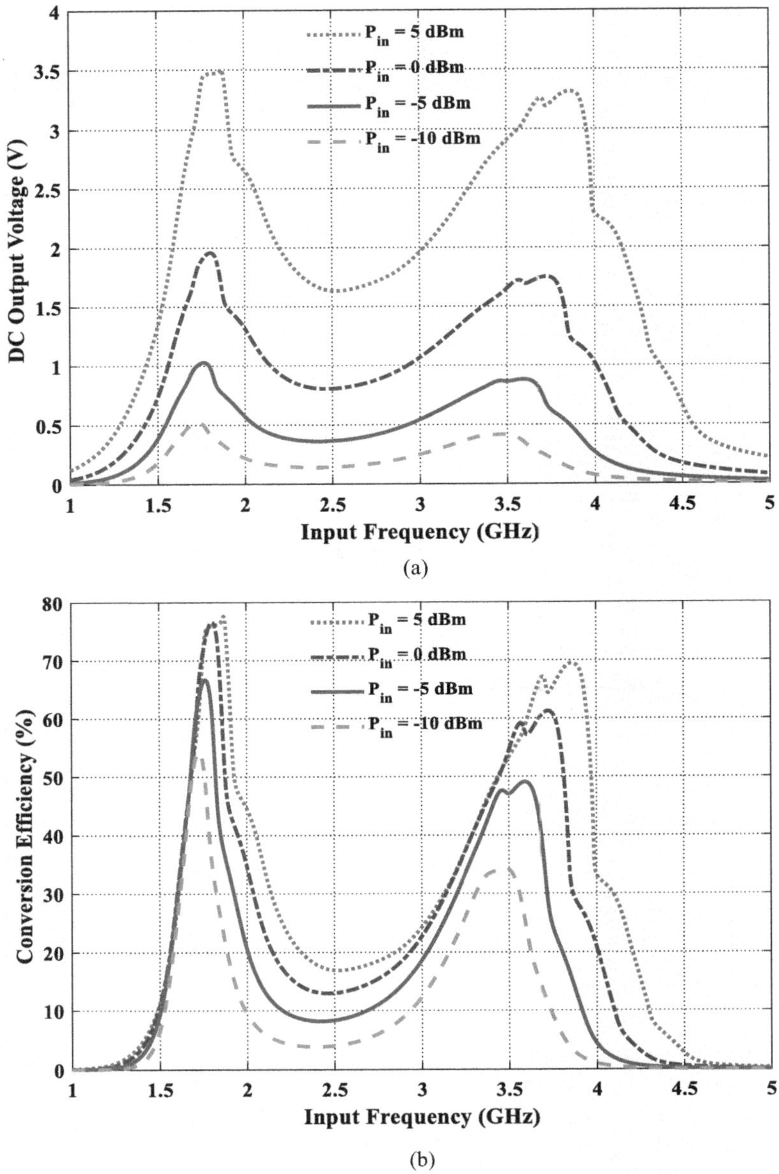

Fig. 11 (**a**) Simulated output voltage, (**b**) PCE with various input power levels at $R_L = 5\ k\Omega$

Table 2 Performance comparison of proposed AC to DC converter with recent work

Refs.	AC to DC converter topology	Frequency (GHz)	Diode used	Impedance matching technique	Pin (dBm)	Output voltage (V)	R_L (kΩ)	Efficiency
[17]	Voltage doubler	0.915 and 2.45	HSMS 2862	Distributed parameter	14.6	–	1	77.2%
						–		73.2%
[18]	Half-wave rectifier	0.9 and 1.8	HSMS25850	Distributed parameter	−30	0.374	4.5	12%
								8%
[19]	Single series diode	0.92 and 2.4	SMS7630	Lumped parameter	0	2.66	4.7	65%
[20]	Full-wave rectifier	2.45 and 5.5	HSMS-2852	Distributed parameter	0	–	12	36%
								8%
This work	**Voltage doubler**	**1.8 and 3.6**	**HSMS 2850**	**Lumped parameter**	**0**	**1.85**	**5**	**76%**
						1.6		**57%**

References

1. Chen W, Zhao S, Shi Q, Zhang R (2020) Resonant beam charging- powered UAV-assisted sensing data collection. IEEE Trans Veh Technol 69(1):1086–1090
2. Zhang X, Zhang X, Han L (2019) An energy efficient internet of things network using restart artificial bee colony and wireless power transfer. IEEE Access 7:12686–12695
3. Divakaran SK, Krishna DD (2019) RF energy harvesting systems: An overview and design issues. Int J RF Microw Comput-Aided Eng 29(1):e21633
4. Wang X et al (2022) A novel high-sensitivity broadband rectifier for ambient RF energy harvesting. Radioengineering 31(3):331
5. Saeed W et al (2018) RF energy harvesting for ubiquitous, zero power wireless sensors. Int J Antennas Propag 2018:1
6. Nagaveni S et al (2020) Resistance compression dual-band differential CMOS RF energy harvester under modulated signal excitation. IEEE Trans Circuits Syst I Regul Pap 67(11): 4053–4062
7. Brown WC (2016) The microwave powered helicopter. J Microw Power 1(1):1–20
8. Valenta CR, Durgin GD (2014) Harvesting wireless power: survey of energy-harvester conversion efficiency in far-field, wireless power transfer systems. IEEE Microw Mag 15(4): 108–120
9. Roy S et al (2021) Quad-band multiport rectenna for RF energy harvesting in ambient environment. IEEE Access 9:77464–77481
10. Almansouri AS, Kosel J, Salama KN (2020) A dual-mode nested rectifier for ambient wireless powering in CMOS technology. IEEE Trans Microw Theory Techn. 68(5):1754–1762
11. Liu J, Zhang XY (2018) Compact triple-band rectifier for ambient RF energy harvesting application. IEEE Access 6:19018–19024
12. Costantine J, Eid A, Abdallah M, Tawk Y, Ramadan AH (2017) A load independent tapered RF harvester. IEEE Microw Wireless Compon Lett 27(10):933–935
13. Zheng S, Liu W, Pan Y (2019) Design of an ultra-wideband high- efficiency rectifier for wireless power transmission and harvesting applications. IEEE Trans Ind Informat 15(6): 3334–3342
14. Song C, Huang Y, Zhou J, Zhang J, Yuan S, Carter P (2015) A high- efficiency broadband rectenna for ambient wireless energy harvesting. IEEE Trans Antennas Propag 63(8): 3486–3495
15. Baroudi U (2019) Management of RF energy harvesting: a survey. In: 16th International Multi-Conference on Systems, Signals & Devices (SSD), pp 44–49, doi: https://doi.org/10.1109/SSD.2019.8893169
16. Wang X, Mortazawi A (2019) Rectifier array with adaptive power distribution for wide dynamic range RF-DC conversion. IEEE Trans Microw Theory Technol 67(1):392–401
17. Liu J, Zhang XY, Yang C-L (2018) Analysis and design of dual-band rectifier using novel matching network. IEEE Trans Circ Syst II Exp Briefs 65(4):431–435
18. Muhammad S, Tiang JJ, Wong SK, Smida A, Ghayoula R, Iqbal A (2021) A dual-band ambient energy harvesting rectenna design for wireless power communications. IEEE Access 9:99944–99953
19. Mansour MM, Torigoe S, Yamamoto S, Kanaya H (2021) Compact and simple high-efficient dual-band RF-DC rectifier for wireless electromagnetic energy harvesting. Electronics 10(15): 1764
20. Mattsson M, Kolitsidas CI, Jonsson BLG (2018) Dual-band dual-polarized full-wave rectenna based on differential field sampling. IEEE Antennas Wirel Propag Lett 17(6):956–959
21. Chen M-C, Sun T-W, Tsai T-H (2022) Dual-domain maximum power tracking for multi-input RF energy harvesting with a reconfigurable rectifier Array. Energies 15(6):2068
22. Wang Y et al (2022) A high-efficiency and reconfigurable rectenna array for dynamic output DC power control. Front Phys 10:150

23. Sharawi MS (2013) Printed multi-band MIMO antenna systems and their performance metrics. IEEE Antennas Propag Mag 55(5):218–232
24. Tiwari RN et al (2019) Neutralization technique based two and four port high isolation MIMO antennas for UWB communication. AEU Int J Electron Commun 110:152828
25. Gayatri T et al (2022) High isolation four-port wrench shaped compact UWB MIMO antenna for 3.1–10.6 GHz band. Prog Electromagn Res C 122:67–82
26. Andrade-González EA et al (2021) UWB four ports MIMO antenna based on inscribed Fibonacci circles. J Electromagn Appl 35:1–19
27. Aggarwal I et al (2022) Design and analysis of metamaterial-based SWB-MIMO antenna. IETE J Res:1–14. https://doi.org/10.1080/03772063.2021.2021820
28. Rajkumar S, Anto Amala A, Selvan KT (2019) Isolation improvement of UWB MIMO antenna utilising molecule fractal structure. Electron Lett 55(10):576–579
29. Wu Y et al Design of a tri-band multiple input multiple output antenna with high isolation for 5G applications. Int J RF Microw Comput-Aided Eng 32:e23153
30. Nazmul HM, Chu S, Bashir S (2019) A DGS monopole antenna loaded with U-shape stub for UWB MIMO applications. Microw Opt Technol Lett 61(9):2141–2149
31. Keshri PK, Sahu SK, Chandel R (2021) CSRR-loaded compact quad port MIMO diversity antenna for UWB applications. IETE J Res:1–11. https://doi.org/10.1080/03772063.2021.2006804
32. Tang Z et al (2019) Compact UWB-MIMO antenna with high isolation and triple band-notched characteristics. IEEE Access 7:19856–19865
33. Sadineni RB, Dinesha PG (2022) Design of Penta-Band Notched UWB MIMO antenna for diverse wireless applications. Prog Electromagn Res M 107:35–49
34. Kumar S et al (2020) A compact four-port UWB MIMO antenna with connected ground and wide axial ratio bandwidth. Int J Microw Wirel Technol 12(1):75–85
35. Choudhary V, Meshram MK, Hesselbarth J (2022) Four elements reconfigurable MIMO antenna for dual band applications. Int J Adv Microw Technol 7(1):274–282
36. Nella A et al (2022) A reconfigurable integrated 4-port UWB and NB antenna system for cognitive radio application. Int J RF Microw Comput Aided Eng 32(3):e22998
37. Naktong W, Ruengwaree A (2020) Four-port rectangular monopole antenna for UWB-MIMO applications. Prog Electromagn Res 87:19–38
38. Saurabh A, Kumar PS, Rathore, and M. Kumar Meshram. (2020) Compact wideband four-element MIMO antenna with high isolation. Electron Lett 56(3):117–119
39. Srinivasu G et al (2021) Influence of FR4 material substrate on diamond slotted ultra wideband antenna in 1.71 GHz to 12 GHz. Mater Today Proc 45:5642–5648
40. Tirado-Méndez JA et al (2020) Four ports wideband drop-shaped slot antenna for MIMO applications. J Electromagn Waves Appl 34(9):1159–1179

Energy Harvesting Systems for Agricultural Needs

Swapnaja K. Jadhav and R. Shreelavaniya

1 Introduction

Energy is an inevitable factor to accomplish any operation. Agriculture is the main source of the food energy for human beings, and therefore agriculture sector is one of the main key developments along with energy sector to flourish the economic sector of any country. Due to limited sources of conventional energy, the focus is on new renewable energy sources and alternate energy production from freely available natural resources. There is ever-increasing demand of food to feed increasing population which is to be met through utilization of limited resources of land, water, manpower, seeds, fertilizers, farming operations, chemicals, etc. Water scarcity, reduced manpower, environmental issues related to chemical use, cost of fertilizer, pesticides, low efficiency of the farm machineries, and loss of direct and indirect energy are the challenges for modern agriculture.

These challenges are the driving key for the smart farming applications which can be achieved through precision agriculture, automation, remote sensing, energy self-sufficient agriculture, etc. Also, urgent establishment of a strategic shift to agricultural sustainability is required. Artificial intelligence and Internet of things (IoT) are two building blocks for the smart agriculture. Abiotic and biotic factors are being assessed through remote sensing and sensors to optimize crop and livestock management [1]. In addition, artificial intelligence implementation and applications have enormous advantages that can revolutionize the agri-food sector and its associated business. Sensors can deploy artificial intelligence in the irrigation process and water

S. K. Jadhav (✉)
ICAR-Central Institute of Agricultural Engineering, Bhopal, India

R. Shreelavaniya
ICAR Research Centre on Bioconversion Technology, Sri Parasakthi College for Women, Tenkasi, TN, India

A. Nella et al. (eds.), *Energy Harvesting Trends for Low Power Compact Electronic Devices*, EAI/Springer Innovations in Communication and Computing,
https://doi.org/10.1007/978-3-031-35965-1_6

management by establishing smart irrigation systems [2] with an understanding of "more crop per drop" along with identification, control, monitoring system, and treatment of sensor-based pests. To overcome the physical labor required in agriculture, precision agricultural equipment like smart tractors, robots for harvesting, agri-bots, unmanned aerial vehicles, wireless technology, etc. are being developed [3]. Postharvest management of agricultural produce is another important area for applications of artificial intelligence. Thus, the intervention of artificial intelligence in mechanization can not only make proper use of resources but also make the hard work easier by helping the farmers save value and time. Smart agricultural applications are difficult due to their environmental applications and sensors involved. However, a crucial part is an energy source to operate the hardware and microprocessor involved in all these technological applications. All the low-energy devices such as sensors and embedded devices need to be powered by batteries. However, use of batteries comes with limitations which can turn out costly.

Energy harvesting is a process of extracting and converting residual energy into electrical energy to operate electronic components like microprocessor. It is also necessary for the applications which require a backup energy source. The energy can be generated from various sources which otherwise get wasted. This energy generally exists as by-product of a natural phenomenon or process, viz., mechanical energy because of vibration, stress, and strain of material [4]; thermal energy exhausted from furnaces, combustion engines, and other heating sources; energy coming from all types of light sources including solar energy; wind and fluid energy coming from air and liquid flow; coils and transformers; electromagnetic energy captured by inductors; chemical energy from naturally recurring or biological processes; and RF energy due to radio transmitters. Energy harvesting systems enabled by new circuits can harness and store small amount of energy and convert them into usable form.

The circuits are designed to broaden the range of voltage, waveform inputs, and current and to tolerate overvoltage, overcharge, and other kinds of irregular input conditions. Various types of sensors are used to convert other form of energy to the electrical energy that can be harnessed, stored, and processed for various low-voltage applications with the help of energy harvesting systems [5]. Energy harvesting electronics have advances in capturing; storing in a capacitor, supercapacitor, or battery; and achieving high efficiency [6].

This chapter will provide brief insight on the energy harvesting system technology application for agricultural need and application of different types of energy harvesting systems for sensor application in agriculture. It will also contribute in the identification of the AI approach for understanding the processes of service creation and decision-making as support for various agri-food applications and supply chain phases.

2 Energy in Production Agriculture

The agriculture sector has major implications for the overall demand and supply of energy. Farm labor and animal energy needs of agriculture have shifted toward mechanical energy. Almost all countries which make significant contribution to world production of food and other crops depend to different degrees on fossil fuels for fulfilling the energy needs of agriculture, not only for mechanical operations like tillage, harvesting, etc. but also for inputs like seed, water, fertilizer, chemicals, and equipment. There is a similar increase in the number of irrigation pumps which is reflected in increase in consumption of electricity. To ensure that production also increases, additional supply of energy in very large quantities will need to be made to production agriculture [7].

2.1 Direct Energy Sources

Direct energy sources for agriculture refer to the energy that are used for activities like land preparation, sowing/planting, irrigation, intercultural, pest control, harvesting, threshing, and transportation of agricultural input resources and farm produce in crop production.

Major energy consumption in the agriculture sector is attributable to irrigation. Since the rapid urbanization taking over the agricultural land, crop production needs to be increased by crop intensifying through raising productivity factors like water, fertilizers, farm mechanization and automation, etc. The sector is a major consumer of diesel in India for tractors and pumps; however, diesel utilization in pumps will decrease due to more preference to electric pumps [8]. Electromechanical power for stationary operations like water pumping, operating power threshers, cleaners, and other on farm processing equipment.

2.2 Utilization of Electric Energy in Agriculture

Electric energy is required in agriculture to operate irrigation pumps, light tillage operations, spraying, threshing, winnowing, shredding, grinding, hot air drying, processing, thermal application, and almost every postharvest operation and their control systems. The electricity demand in agriculture mainly depends on the amount of water to be pumped per year. However, irrigation requirements are increasing day by day due to uneven rainfalls and increase in cropping intensity. Hence, reliable irrigation is a critical requirement not just for the farmers but also for the country. The source of irrigation has shifted dramatically to groundwater over the last two decades, with little regulation of well capacity or aquifer yield assessments [9].

It is necessary to have a reliable energy source to make the equipment, machineries, and electronic components functional. The self-sufficient energy harvesting can be classified into macro-applications and micro-applications. Macro in the sense that energy generation can be carried out using resources available on farm which can be either utilized to run the farm equipment, irrigation pumps, postharvest operations, etc. or stored in the batteries for other low-energy applications like precision agriculture, remote sensing, and IoT-based monitoring systems. This includes solar photovoltaics (PV), biomass gasification-based power generation, biogas-based power generation, etc. Micro-application includes the energy generated with the help of embedded hardware for microprocessors using active sensors and storage technologies which can be used again in smart agricultural application using AI and IoT.

3 Smart Farming

A modern agriculture required wireless sensor network broadly termed as IoT. IoT-based agriculture can be used for monitoring the crop growth dynamics with the help of sensors (light, humidity, temperature, soil moisture, nutrient profile, etc.), storage facilities, livestock monitoring, smart agriculture vehicles, autonomous robots, drones, smart greenhouses, hydroponics, predictive modeling, and planning and irrigation scheduling system. The farmers can monitor the field and livestock and get warnings regarding desired output or emergency from remote places. In agriculture, there is huge potential of sensors and communication systems based on wireless nano-sensors which requires sufficient and constant source of power to keep them running efficiently [10, 11].

The energy sources are external energy source like battery and direct solar photovoltaics or any other stand-alone or grid-connected energy supply. For agriculture, energy generation technologies are explained in Sect. 5. There are various types of energy harvesting systems used along with sensors. Figure 1 shows various applications of energy harvesting systems required in agriculture.

4 Pathways of Energy Supply in Advance Agriculture

The macro-energy produced of high voltage and current generated from wind and biomass-based power plants can be used mainly in two pathways as main power supply and power supply to battery and super-capacitors.

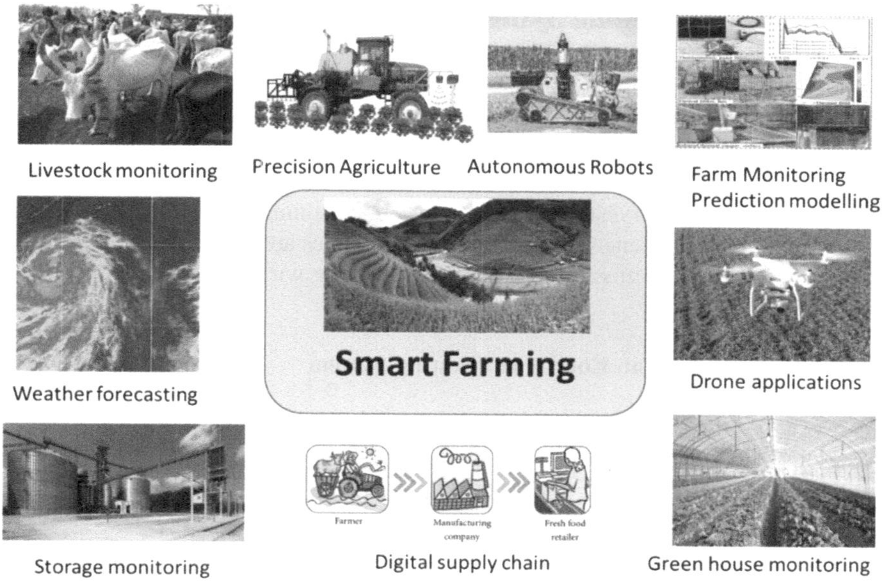

Fig. 1 Energy harvesting system in smart farming applications

4.1 Main Power Supply

The energy generated from solar photovoltaic plants and some of the biomass energy harvesting system like gasification power plants, biogas power plants, heat pump, etc. is suitable for major mechanical operations in the agriculture. This is more suitable for IoT or remote applications with stationery devices, where a power supply can be connected through wires or cables. However, this may make the devices stationary and restricts its applications with large number of devices.

4.2 Power Supply Through Battery and Super-Capacitors

Battery is the typical portable energy source which has been usually and widely used in various devices. The main power supply can be used to charge the batteries and then used for the various applications. However, there is limitation regarding time duration and the life of the battery. Batteries become expensive to use to maintain and replace over a particular time duration in IoT or remote sensing applications because the sensor nodes may be widely distributed over a remote area. Non-rechargeable batteries also have a restriction due to environmental implications and their limited storage capacity. However, sensors work at ultralow power, which can be assisted by different sizes of batteries, and hence the batteries can play a major role in some IoT applications [12].

4.3 *Energy Harvesting Systems for AI and IoT Devices*

Conventionally for low-power electronics, such embedded and remote sensors are powered by batteries [13]. To avoid the expensive cost, maintenance cost, and limited operating life, energy harvesting systems can play a better role because energy harvesting technologies provide lifelong operation to low-power devices. Many of the energy harvesting systems are self-sustaining, are cost-effective, and require minimum maintenance. Since the power sources are attached with the sensor nodes, it excludes transmission losses and avoids long wires and cables.

4.3.1 Basic Parts of an Energy Harvesting System

Energy harvesting system needs four major units for conversion of the energy input in various forms as shown in Fig. 2:

(i) *Source of energy*: Different energy sources as motion, heat, light, vibration, radio frequency, etc. as an input.
(ii) *Transducer/converter*: These are the energy harvesters which collect and convert the form of energy into electric energy. Different types of transducers convert the energy of light using photovoltaic, magnetic using inductive, heat using thermoelectric, radio frequency using RF energy harvester, and vibrations or kinetic energy using piezoelectric or electromagnetic transducers.
(iii) *Energy storage*: Super-capacitor or small chargeable battery.
(iv) *Power management component*: This unit converts and stores input energy into a suitable form of energy. This includes regulators and complex control circuits that can manage the power, based on requirement and the available power.

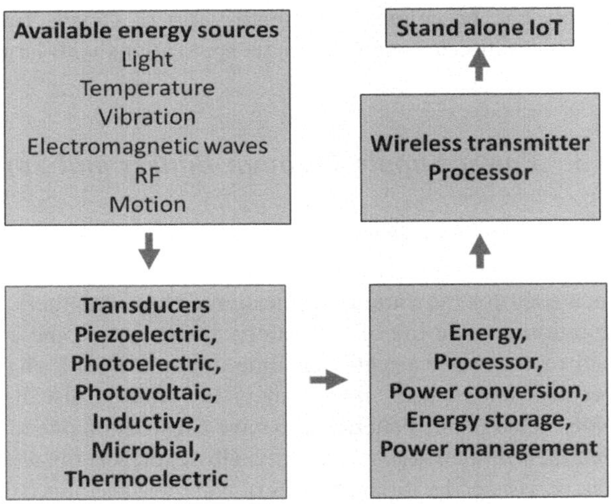

Fig. 2 Overview of the energy harvesting system

Energy harvesting systems require power conversion circuit for efficiently collecting, converting, and managing the energy from these sources into required power for wireless devices, microcontrollers, sensors, and other low-power devices.

5 Energy Generation Systems for Main Power Supply

5.1 *Solar Energy Harvesting Through Solar Photovoltaic Technology*

Solar energy can be directly converted into electrical energy. Energy conversion devices which convert sunlight into electrical energy by use of the photovoltaic effect are termed as a solar cell.

5.1.1 Agri-Voltaic Systems

Agri-voltaic systems are PV-based systems in which PV panels are mounted at a sufficient height from the ground to enable the crop cultivation. These systems preserve agricultural land while generating power and benefit crop production by improving water use efficiency and reducing water stress [14]. Agri-voltaic systems provide a potential solution to the land economy problem.

5.2 *Biomass Energy Harvesting System*

Organic material produced by plants, including terrestrial (those grown on land), aquatic (those grown in water), and its derivatives, is referred to as biomass. It comprises crops and residues from forests, farms, and energy plantation specifically for their energy value and animal manures. Biomass can be regarded as a renewable energy source because plant life renews and grows every year, unlike conventional fossil fuels, which take millions of years to form [15].

5.2.1 Types of Biomass Available from Agriculture

The biomass resources can be classified into various criteria. It might be either originated from plants or animals. It can be animal manures; biogas slurries; crop residues; energy crops; organic wastes; by-products and wastes from agro-industries, food industries, and fodder and brewery industries; wastewaters and sludges from agro-industrial processes; wastes from biofuel production and biorefineries; etc. [16, 17].

The agriculture sector provides most of the feedstock substrates used for biogas and has the greatest potential as a source of biogas feedstocks. These feedstocks are mostly made up of different wastes and by-products, with farm animal manures and slurries being the most significant such as cattle dung, pig's excreta, poultry droppings, etc. Before feeding them into a digester, most of them need to be pretreated. Pretreatments can be as simple as mechanical particle size reduction [18] or as complex as breaking down the lignocellulosic molecules to make it easier for anaerobic microbes to access these structures. Crops such as maize, grasses, beets, and sunflowers are grown particularly to produce biogas.

5.2.2 Crop Residue Availability

Approximately 50 EJ, or 10% of the world's yearly primary energy consumption, is provided by biomass, which is widely employed for energy purposes [19]. According to the World Energy Council, there is still a great amount for growth in the use of biomass for energy, including the production of biogas.

5.2.3 Biomass Gasification and Power Generation

Gasification is the process of conversion of biomass into combustible gas by supplying restricted amount of oxygen. The product obtained during low-temperature gasification process (< 1000 °C) is the producer gas, which contains CO, H_2, CH_4, and hydrocarbon. The product obtained during high-temperature gasification (> 1200 °C) is syngas, which contains CO and H_2. Syngas or producer gas can be burned to generate energy in the form of heat, steam, or electricity.

Gas engine can be run on 100% producer gas. Biomass required to produce 1 kWh of power from gas engine is varied from 1.2 to 1.5 kg. Approximately, 70–80% diesel can be replaced using producer gas by running dual-fuel engine. Biomass and diesel fuel required to produce 1 unit of electricity are 0.9–1.2 kg and less than 0.1 L, respectively [20].

5.2.4 Biogas-Based Power Generation

Biogas contains high amount of methane, and it is used as an excellent fuel for internal combustion engines. The ignition temperature of biogas is 640–840 °C [21]. Biogas contains the mixture of methane and carbon dioxide, which mixed with air during combustion that runs the engines producing the power. The use of this biogas in internal combustion engines saves huge amount of fossil fuels. 1 hp. engine requires 1 m^3 biogas to work for about 2 h roughly. Biogas can be used in both CI and SI engines [22]. CI engines run on dual-fuel mode (biogas and diesel), and injection of diesel is necessary for igniting the mixture of biogas and air along with initial start-up that is done with pure diesel alone. The CI engines can be converted to use biogas up to 80% replacement of diesel with slight modification. Thereby, it reduces the diesel oil consumption. SI engines can be operated using biogas initial start-up that is done with petrol fuel. The amount of gas required for producing 1 hp. power in 1 h is 0.425 m^3. Biogas can be used in SI engines with slight modification in carburetor [23].

5.3 *Small Hydro Energy Harvesting Systems*

Hydropower is electricity produced by the mechanical movement of water from rivers and lakes [24]. Gravity cause water to move downward, and this downward motion of water contains kinetic energy, which can be converted into mechanical energy and then into electrical energy.

Small hydro can be classified into three categories:

(a) Micro (less than 100 kW).
(b) Mini (100 kW to <1 MW).
(c) Small (1 MW to <10 MW).

Small hydro systems operate by diverting part of the river flow through a turbine, which activates a generator to produce electricity [25].

5.4 *Solar Heat Pump*

The heat pump technique is based on mechanical energy conversion, which provides heating or cooling by using a reversible refrigeration cycle [26]. Essentially a simple heat pump consists of:

(a) Compressor.
(b) Condenser.
(c) Evaporator.
(d) Working fluid (Freon).
(e) Expander.

The solar-assisted heat pump works in the gaseous state, which is compressed and sends to the condenser to get cooled to liquefy and releases large amount of heat to the surrounding. Then, through expansion valve, this liquefied working fluid gets vaporized. The evaporator requires large amount of low-temperature heat, which is supplied by solar-heated air or water [27]. Using valve and controls, the heating in cold weather and cooling in hot season can be provided by the same unit.

5.5 *Fuel Cells*

Fuel cell is an electrochemical device that converts chemical energy of a reaction into electrical energy. In a typical fuel cell, gaseous fuels are fed continuously to the electrode compartment. The electrochemical reactions take place at the electrodes to produce an electric current. The most common classification of fuel cells is as follows.

5.5.1 Alkaline Fuel Cell

The electrolyte of this fuel cell is concentrated with KOH (85%) in the fuel cells operated at temperature of 250 °C or KOH (30–50%) at lower temperature less than 120 °C [28]. Asbestos matrix and a wide range of electrocatalysts can be used (e.g., Ni, Ag, metal oxide, and noble metals) in this fuel cell.

5.5.2 Molten Carbonate Fuel Cell

In this fuel cell, a combination of alkali carbonates is used as electrolyte in a ceramic matrix of $LiAlO_2$. The alkali carbonates form a highly conductive molten salt with carbonate ions at 600–700 °C providing ionic conduction [29].

5.5.3 Phosphoric Acid Fuel Cell

Phosphoric acid 100% (conc.) is used as the electrolyte which operates at 150–220 ° C in this type of fuel cell.

5.5.4 Polymer Electrolyte Fuel Cell

This fuel cell has an ion exchange membrane (fluorinated sulfonic acid polymer) as an electrolyte which is an excellent proton conductor.

5.5.5 Tubular Solid Oxide Fuel Cell

This fuel cell contains a solid electrolyte of nonporous metal oxide, usually Y_2O_3-stabilized ZrO_2. Ionic conduction by oxygen ions takes place at 1000 °C. Usually, the anode is Co-ZrO_2 or Ni-ZrO_2 cermet, and the cathode is Sr-doped $LaMnO_3$.

5.5.6 Intermediate-Temperature Solid Oxide Fuel Cell

The materials in this fuel cell have the same materials used for electrode and electrolyte as that of tubular solid oxide fuel cell. This fuel cell operates at temperature in the range of 600–800 °C [30].

6 Types of Energy Harvesting Technologies for Agricultural Application

Various types of energy harvesting technologies which have diverse range of applications in the agricultures are described in this section. These energy harvesting systems are necessary for wireless sensors to function. The energy available in various forms can be captured by the system based on suitable energy harvesting phenomena. Some available technologies necessary for use in the wireless sensors are described [31].

The energy can be available in the form of vibrations of machinery, condensers, body movement or body dynamics of humans or animals, moisture, solar energy, available RF energy, body heat, external heat from the microbial activity, environmental conditions, due to evapotranspiration, temperature difference, body motions, air velocity, etc. various energy harvesting technologies, viz.:

(i) Piezoelectric energy harvesting.
(ii) Thermoelectric energy harvesting.
(iii) Pyroelectric generator.
(iv) Radio frequency harvester.
(v) Electromagnetic.
(vi) Electrostatic.

6.1 Piezoelectric Energy Harvesting

Piezoelectric energy harvesting is based on the energy available in the form of mechanical energy. This mechanical energy can be identified in the form of vibrations, pressure, and stress-strain. Piezoelectric mechanisms generate electricity from mechanical energy. There are piezoelectric materials in which strain induced by vibrations or pressure which induce an electric potential at the terminals of a piezoelectric material. Figure 3 shows the operating principle of piezoelectric effect. The energy can be generated with the displacement due to stress and strain which create the flow of charge in direct and reverse direction.

Mechanical energy sources for application of piezoelectric harvester operation are vibration due to livestock body movement [32], displacement due to water level, air pressure inside the storage bins, etc. Piezoelectric materials produce high power

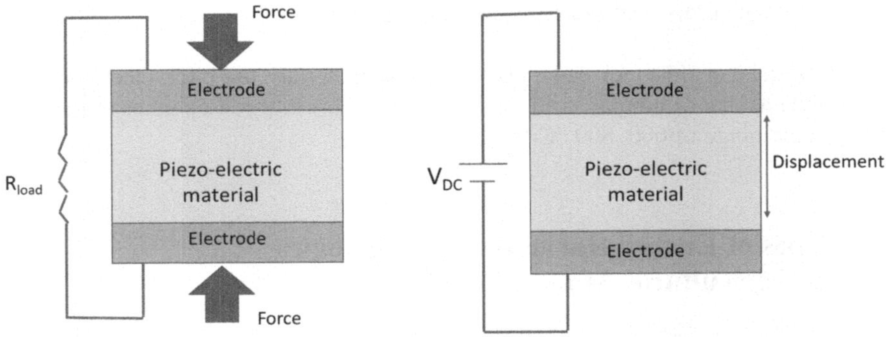

Fig. 3 Operating principle of piezoelectric effect. (**a**) Direct. (**b**) Reverse/converse

density, and advanced new materials with enhanced properties make them very suitable for the applications in Internet of things.

6.1.1 Piezoelectric Animal/Human Motion Energy Harvesting

Draft animals or human bodies generate a significant amount of energy while working in agricultural field. Harvesting this energy can overcome battery capacity limitation for long-duration operation of wearable devices for monitoring. Wu et al. [33] integrated a spring pendulum device with a piezoelectric ceramic. This device has 2 degrees of freedom due to pendulum and shows that the harvester can generate 5.80 mW electrical power under the excitation of common human motion signals (1.75 Hz, 0.20 g). This device with custom design can be applicable to the livestock with some damping arrangement since animal movement is quite uncertain [34].

6.1.2 Energy Generation from Animal Body

Studies have been carried out on implantable acoustic micro-transmitter which is solely powered by a biomechanical energy harvester and was demonstrated in vivo in fish and under quasi-realistic conditions. A piezoelectric macro-fiber composite beam was used to harvest the energy generated from a fish's body. This is one alternative pathway to achieve an efficacy of powering implantable and wearable electronic devices in animal body [35].

6.2 *Thermoelectric Energy Harvesting*

In thermoelectric systems, electric voltage or electric current can be generated due to temperature differences across the material [36, 37]. The voltage can be maintained up to 5 V based on the temperature difference. Seebeck effect is the generation of voltage when the change in temperature is sustained between two sides [38, 39]. Thermoelectric generator (TEG) is a device for electric generation which uses semiconductors. The N- and P-doped semiconductor materials are connected electrically in a series and thermally in parallel, to generate higher voltage as shown in Fig. 4. Ceramic layer is used as isolation between electrical and thermal components. A TEG usually has low efficiency in the range of 5–10% which restricts its adoption in application [40]. However, it has a long-life cycle and stationary parts [41].

Soil is the good source to get a temperature difference. The energy collection devices can be developed for the other IoT applications in agriculture. As the depth of the upper layer of the soil increases, the temperature difference between ambient temperature and soil temperature increases. The temperature gradient between the surrounding and the soil is considered for the operation of TEG. To harvest the energy from this phenomenon, collection device needs to be developed attached with TEG. This device needs to embed into ground [42].

Higher moisture content of soil can generate higher voltage due to thermal conductivity. The voltage generated depends upon types of soil, surrounding temperature, the weather, sunlight intensity, the landscape of agriculture, vegetation, etc. The care should be taken to avoid the moisture loss of soil, vegetation to keep soil at lower temperature, and physical objects surrounding the device blocking the direct sunlight dropping on the ground. Higher temperature difference without any disturbance such as wind from the surrounding or nature will produce more electrical energy from the TEG Aridi et al. [43].

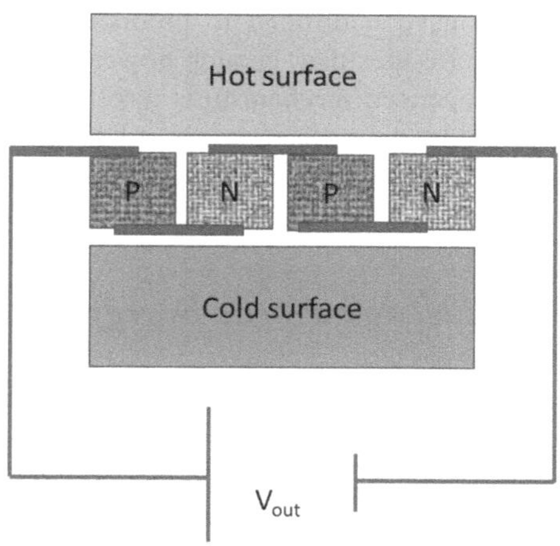

Fig. 4 Schematic of thermoelectric generator device

6.3 Electrostatic Energy Harvesting

The charged capacitor plates of electrostatic converter are displaced using vibration; the deviation in capacitance makes it convert into electrical energy. Surface charges will then pass from the capacitor to a storage device or to the load. In this case, the mechanical vibration motion between two plates is converted to electrical energy. Electrostatic energy harvesters can be classified into three types as in-plane overlap, which varies the overlap area between electrodes; in-plane gap closing, which varies the gap between electrodes; and out-of-plane gap closing, which varies the gap between two large electrode plates [44].

6.3.1 Animal Monitoring Using a Combination of Triboelectric and Electrostatic Converter for Biomechanical Energy Harvesting

Energy harvesting devices based on combination of the conversion mechanism can have a high output and high conversion efficiency. An electromagnetic energy harvester coupled with a triboelectric mechanism for biomechanical energy harvesting is studied by Zhai et al. [45] . The output was improved, and a sustainable and enhanced peak power of about 289 μW was produced at the external impedance of 100 MΩ, which is about 46 times the conventional single triboelectric conversion mechanism. It can constantly monitor various motion states. Furthermore, it can charge a capacitor of 10 μF to 3 V and sufficient for 16 LEDs. This can be used in wearable device for humans and animals [46].

6.3.2 Precision Planters or Speeders

The vibrations can be detected and the positioning of the machine components needs to be carried out. However, variations in the vibrations due to the undulating land profile make it difficult to produce controlled vibration input to the harvesting system. Besides that, enough power is available through prime movers, and hence power-operated mechanism is mostly being used instead of sensors. Farmers with big land holding use sensor-based precision planters, and the use may spread once the awareness and skills required to operate such machinery in farmers increased.

6.4 Electromagnetic Energy Harvesting System (EEHS)

Electromagnetic energy harvesting systems based on vibration provide a renewable energy-based cost-effective source. Electromagnetic vibration energy harvesting is a renewable alternative to batteries as a power source for IoT and wireless sensor nodes. With the smart farming development, autonomous agricultural machineries equipped with various sensors are of inevitable part of it. Vibration is one of the by-products of the operating machineries. In this system, mechanical-to-electrical energy conversion is carried out with overall efficiency of about 30% [47].

6.4.1 Sensor-Based Automated Agricultural Machineries

Fixed frequency vibration, sinusoidal sweeping vibration, and stochastic road spectrum tests of agricultural machinery can be carried out to know the power generation capacity of a developed vibration-based energy harvesting system for particular agricultural machinery. Peak voltage and power generated from such devices are about 44 V and 113 mW under low frequency and high displacement excitations which is in the range of the sensors used in agricultural production. This technology is mostly used in the weeding robots, spraying robots, automated tractors, self-propelled sprayers, transplanters, weeders, etc.

6.4.2 Energy Harvesting from the Motion of Animals

Animal monitoring sensor devices to monitor temperature, pulse, etc. need a very little energy to operate. The EEHS as a wearable device are suitable to monitor the animals while performing the operation [48].

6.4.3 Electromagnetic Wind Energy Harvester

Modern agriculture operates on the autonomous Internet of things (IoT) for potential agricultural application. Wind energy harvesting system for IoT-based environment monitoring is most feasible for the agriculture. A wind-driven electromagnetic energy harvester using rotational magnet rotor paired with a back-iron shield shows enhanced magnetic flux density on the front side of a rotor where the connected coils interact and convert the wind energy into electrical energy through electromagnetic induction. A well-designed power management circuit along with the energy harvesting systems gives real-time environmental monitoring. Such type of harvester can generate a maximum average power of 1.02 mW and maximum power efficiency of 73% while operated at about 4.5 m/s wind speed. This wind-driven energy harvesting system can power a commercial wireless sensor that transmits temperature and humidity data to other devices for more than 3 h after each charging. The experimental study recommends this wind-driven system can be implemented in IoT for smart farming applications Roy [49].

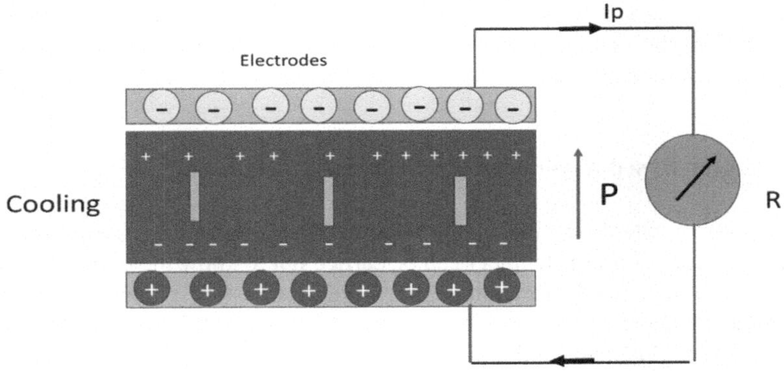

Fig. 5 Mechanism of the pyroelectric energy harvesting system

6.5 *Pyroelectric Energy Harvesting*

This system converts change in temperature into equivalent electric charge. When the temperature of crystal material changes, it induces electric charge as shown in Fig. 5. No charge is generated when temperature remains constant. This method is used in certain sensors. Satyanarayan Patel and Nikola Novak [50] studied a ferromagnetic material $Pb_{0.99}Nb_{0.02}[(Zr_{0.57}Sn_{0.43})_{0.92}Ti_{0.08}]_{0.98}O_3$ (PNZST) for high energy density harvesting and storage. The Olsen cycle is performed on the ferromagnetic materials for waste heat-to-electrical energy conversion. They obtained energy harvesting density of 1.0 MJ/m^3 per cycle at temperatures of 29.85 and 129.85 °C passing through the electric field between 1 and 9 kV/mm. The high energy storage of 0.9 J/m^3 and energy efficiency of 81% were obtained at 129.85 °C under an electric field of 9 kV/mm.

6.5.1 Pyroelectric System Application in Agricultural Machinery Waste Heat Energy Harvesting

The conversion efficiency depends upon the temperature of a waste heat source [51]. Agricultural prime movers like tractors, self-propelled transplanter, reaper binders, threshers, and combine harvester are mostly run on diesel engine whose exhaust gases are around the temperature of 70 °C, while the engine parts heat up at higher temperature up to 120 °C. The heavy-duty operations can provide the heat from the moving parts due to friction. Pyroelectric energy harvesting systems are suitable for the energy generation for operating sensors used on such machineries.

6.5.2 Pyroelectric System Application for Postharvest Monitoring

Drying of agricultural produce is a necessary process after the harvesting of the produce. Solar dryers such as cabinet dryers and tunnel dryers are innovated now with the sensor-based interventions. Temperature and humidity sensors of these dryers can run on pyroelectric energy harvesting systems. Once the agriculture product is kept in the dryer, with the increase in temperature, the drying rate increases which also increases the humidity of the dryer. The higher humidity inside the dryer inhibits the drying rate. Therefore, removal of the inside air after attaining the humidity is necessary. For this purpose, most of the dryers are provided with exhaust fans based on the input of the humidity sensor. These sensors can be operated with the help of pyroelectric energy harvesting system.

Sensors for such systems are currently in used using pyroelectric-based energy.

6.5.3 Pyroelectric Energy Harvesting from Agro-Produce Processing Operation

Agricultural produce processing systems are energy-intensive operations. Thermal applications are required in blanching, canning, boiling, drying, baking, etc. Thermal sources used in these processes lose the heat in the form of exhaust gases, and some part of heat is absorbed by the equipment, vessel, or container. The temperature of these gases and container can be in the range of 140–340 °C or more. This temperature can be used for the energy harvesting using pyroelectric converter. It will be useful if scaled up for various applications.

6.6 Radio Frequency (RF) Energy Harvester

RF energy harvesting device consists of controlled transmitter, RF energy, sensor node, antenna, rectifier, and voltage regulator circuit. The radiated RF from a transmitter is captured by an antenna of the sensor node and converted into DC voltage in the desired range through combination of rectifier and voltage regulator circuit which is further stored in an energy storage system to use to operate the sensor as and when needed.

Desired power generation from RF energy harvesting is generally in the milliwatt and microwatt range based on commercially available transmitters and the distance from RF sources.

RF energy harvesters can be of various types depending upon the application. A typical harvester can provide basic signal rectification, and for that external power management circuit is necessary. A complex RF harvester may integrate the power management and other functions within a single component as shown in Fig. 6. The harvester requires high sensitivity in case of ultralow levels of RF energy. The

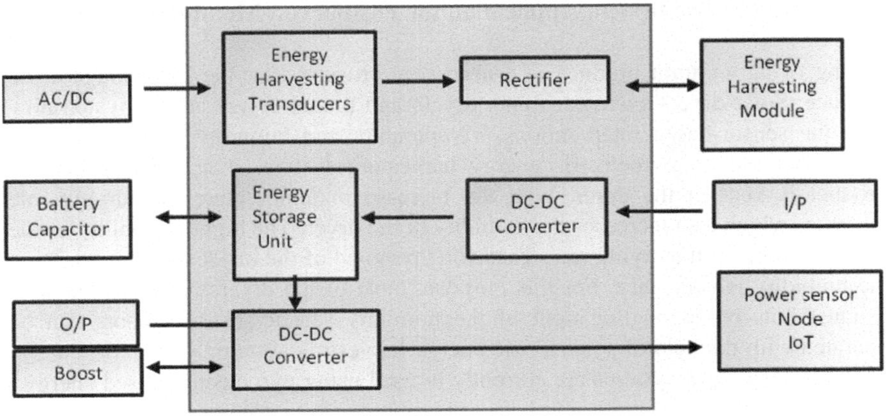

Fig. 6 Typical wireless sensor network with RF energy harvesting system

design of the RF energy harvester should be capable of working efficiently to support a wide range of operating conditions such as input power, output voltage, and load resistance. It should be compatible with commercially available antennas so that it can be implemented to an input impedance of 50 ohms.

RF energy harvesting is a technology that can enable wireless controllable power over distance, scaled up to supply power to thousands of wireless sensor nodes. These devices can be embedded within structures or made portable. This technology can be used in various sensing devices in different applications.

6.6.1 RF Energy Harvester for Pest Detection and Monitoring

RF harvesters can be used in agriculture for various purposes as it serves the wireless sensors. Wireless soil sensor network can be used for in-field pest detection and monitoring. This network consists of numerous sensor nodes which are distributed across an agricultural field. The soil sensor nodes are stationary and can be located anywhere regardless of surrounding conditions. RF energy harvesting provides the power required to operate the soil sensors with more control on power supply compared to other energy harvesting methods. The transmitters can be mounted on a tall structure located at average distance from the sensors [52].

Wireless network of the transmitting sources to control their operation can maximize the performance of a power distribution system. In RF energy harvesting system, operation of power sources as an access point makes it a complete infrastructure for wireless power and data and act as an alternative to the battery. In case of remote farming and greenhouse, this technology is achievable at low cost eliminating the excessive labor cost.

7 Different Energy Sources Available from Agriculture for Energy Harvesting

In Table 1 it is briefly compared between the different energy sources used for different harvesting techniques and the average power output and their applications in agriculture energy harvesting. For IoT applications in agriculture, stand-alone energy harvesters based on different techniques are described explaining their working principles.

Above energy sources are utilized to provide power to the different sensors through energy harvesting systems. Smart farming applications like precision farming, remote sensing, and IoT-based other applications in agriculture required such independent energy sources to operate the sensors.

Table 1 Power generation through different energy sources for energy harvesting systems

	Power	Application in agriculture	
Solar/PV	100 mw/cm^2	Irrigation monitoring, smart gardening, livestock monitoring, pest monitoring	[53–56]
RF energy	40 mw/cm^2	Livestock tracking using wearable RFID tags, animal nutrition monitoring, product tracking, quality monitoring, waste management	[54, 57–59]
Body heat	60 mw/cm^2	Wearables, livestock, labor activity tracker, animal nutrition monitoring, pest monitoring, GPS tracking, long-term body monitoring, emergency detection	[58]
External heat	135 mw/cm^2 at 10 °C	Product tracking, fire detection, food product quality tracking	[31, 60–66]
Body motion	800 mw/cm^3	Livestock activity tracker, livestock nutrition monitoring, GPS tracker, long-term body monitoring, emergency detection	[67–69]
Vibration	4 mw/cm^3	Livestock monitoring, pest monitoring, irrigation monitoring, smart gardening, environment monitoring, water quality monitoring, flood monitoring, hazard detection	[55, 70–78]
Wasted heat of automobile engine	51.13 w/cm^2	Battery charging to run sensor-based operations	[63]
Ferromagnetic material in soft nature	3.61 mw/cm^2	Livestock activity tracker, livestock nutrition monitoring	[58]

8 Application of Energy Harvesting Systems in Smart Agriculture

8.1 Precision Farming

The objective of precision agriculture is to enhance the accuracy of real-time data that can help in decision-making. Precision agriculture saves time and energy involved. The precision farming also gives environmental benefits and improves production process documentation. In coming days, every agricultural farm will be able to use information and technology such as the global positioning system (GPS), geographic information systems (GIS), and navigation systems. The application of various sensors data of the real-time crop, soil and environmental parameters and implement in the operation such as tillage, weeding, sowing, fertilizing, plant protection, harvesting, etc. In many agricultural machineries, optical sensors are used. Energy required for the operation of sensors is derived from the photovoltaic cells, thermoelectric in case of daytime agricultural operations, and battery charged by solar cell or vibration-based electromagnetic energy harvester in case of nighttime operation. Various mechanical and chemical weeding operations use these types of energy harvesters. These sensors-harvesters are capable to detect, record, and measure the radiation of electromagnetic energy and as well as reflected energy. There are different types based on the registered range of Electromagnetic (EM) spectrum, design, detection, recording, measurement, and representation of the detected energy [40]. These are called active sensors because they are coupled with harvester and produce their own energy. Most of the precision devices consist of sensor, energy harvester, a processor, and display for representation. Most of the energy harvesters used are electromagnetic energy harvester because inductive sensors used in these devices are resistant to moisture, dirt, nonmetal parts, and roughness and vibration:

- Photoelectric sensors are also commonly used such as SPAD and the ClorofiLOG sensors. Thermoelectric harvesting systems are used based on the temperature difference between the ambient and soil resulting in electric voltage [79]. This technology is currently in use for crop nutrient and pigment assessment.
- Electrostatic energy harvesters used with capacitive sensors. Change in the distance between surfaces due to impacts is utilized by capacitive sensors to identify the difference in object positioning [80].
- Photoelectric and thermoelectric energy harvesters are used with NIR sensors that are used in detecting plant disease and stresses during the daytime and for the crop protection using the variable rate technology (VRT). With the change in the luminous flux on the sensor based on plant properties, an electronic signal can be passed to the regulator which performs nozzle parameters [81]. The sensor indicates the nitrogen level and plant biomass, and this data is used to process the optimal dose of chemicals in real time.

8.2 Solar Energy-Powered IoT-Based Irrigation Scheduling

IoT is a compulsory part of smart farming which comprises small wireless sensor networks (WSN) used as IoT nodes in agricultural monitoring. Solar energy is used for energy supply for WSN for monitoring and control of agricultural fields [82]. Agricultural monitoring includes crops managements, harvesting, irrigation control, animal control, plant protection control, humidity, temperature monitoring, etc.

Figure 7 shows that the moisture sensor and temperature sensors transmit the data of humidity and temperature in the crops or soil to the remote gateway location. Depending on the processor's algebraic program, the necessary actuators can be activated like motor pump speed or other applications controls. The control can be done from any mobile or computer having required program or application. Such solar-based IoT applications are necessary and suitable for the remote locations like agricultural fields.

8.3 Solar-Based Agricultural Robots

There are some commercial robots available which run on solar energy and use the energy for sensor operations and can be controlled wirelessly from remote place.

"EcoRobotix" robot senses weeds in the field and then moves spray nozzle to the weed to achieve goal of directly delivering herbicide to each weed plant.

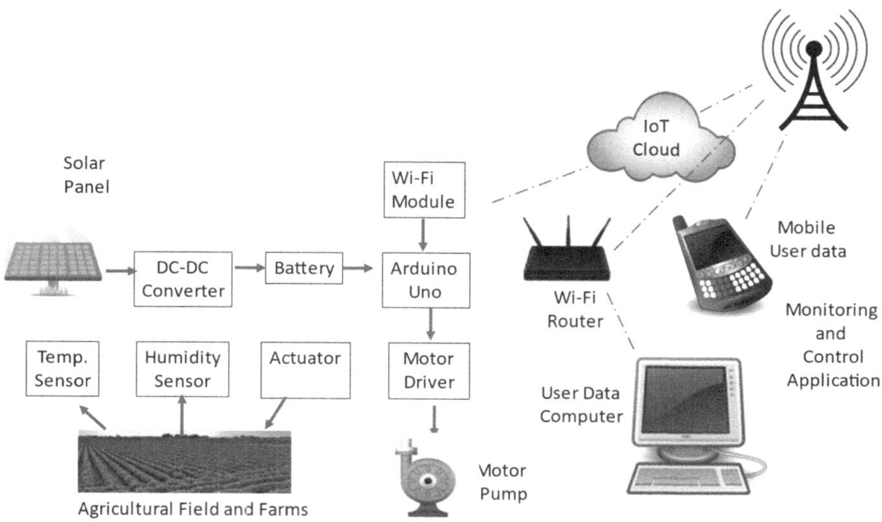

Fig. 7 Use of solar harvesting for IoT-based irrigation scheduling

“Ladybird” robot by University of Sydney’s Australian Centre for Field Robotics features a novel design covered with solar panels and a selective spraying system in which the spray nozzle is moved close to weed plants by a six-axis robotic arm [83].

8.4 *Energy Harvesting Systems and Remote Sensing*

Remote sensing-based agricultural management is done with the help of satellites, aircraft, unmanned aerial vehicles (UAV), and the information they collect. This data can be used and processed for higher-level decisions to manage food supply needs more effectively [84]. Remote sensing requires most necessarily the wireless sensor network or the intermediate transmitters. For which the stand-alone energy supply can be provided by the various types of energy harvesting systems based on the source of energy. The satellites operate on the solar energy harvesting systems. One or more components of remote sensing system require independent energy supply. The significant spatiotemporal challenges can be prevented by using remote sensing technology in various fields. The remote sensing can be used for various applications such as:

- To capture and collect accurate information regarding the boundaries of the cropland area, cropping pattern, etc. which can be useful for the farmers and public agencies.
- Monitoring the large group of livestock using the wearable sensor devices.
- To collect the data regarding types of land such as unproductive soil, wasteland, cropland, etc.
- To detect the plant diseases.
- To detect the pest/animal attacks on the farm such as locusts and wild animals with the help of drones [85]
- To monitor the drought.
- To detect the extreme temperature and the trend over the season.
- Monitoring crop growth parameters, e.g., nutrient deficiency, yield estimation [86]
- To monitor fungal, bacterial, or viral infection in plants.
- To monitor the watershed area over the period.

Based on these information, various decisions can be taken such as requirement of cereals, pulses, or vegetables in coming years, estimation of yield and its supply, draught management, pest management, deployment of control measure facility to the pest infestation area, fertilizer management, etc.

9 Conclusions

Energy harvesting systems have made wireless sensor networks highly suitable at low cost in agricultural field. Energy harvesting also made the portable and wearable livestock health monitoring easy as it eliminates the battery especially based on solar power for the daytime operation integrated with other harvesters. Among the various harvesting systems such as electromagnetic, electrostatic methods and piezoelectric energy harvesting produce power at higher level with higher flexibility. Different energy sources are capable to produce different level of power density, such as 800 mW/cm^3 using body motions and 135 mW/cm^2 using external heat followed by 100 and 800 mW/cm^3 through solar energy and then by other energy sources.

The unstable environmental conditions and the low adaptability of the systems reduce the reliability of the energy harvesting systems for agricultural purpose. Combination of energy harvesting systems can be more efficient and reliable in agricultural field. Energy harvesting plays a major role in remote sensing, precision agriculture, IoT-based monitoring, and controlling of agriculture-related operations and events which facilitate better management and conservation of resources, time, and energy resulting in higher productivity. The low output, less efficiency, and low reliability are challenges that restrict their practical application in some of the areas of agriculture. An integrated approach of energy harvesting may result in the higher efficiency and higher reliability.

References

1. Adhikary S, Biswas B, Naskar MK, Mukherjee B, Singh AP, Atta K (2022) Remote sensing for agricultural applications. In: Eyvaz APM, Albahnasawi A, Gürbulak E, Tekbaş M (eds) Arid environment – perspectives, challenges and management. IntechOpen. https://doi.org/10.5772/intechopen.106876
2. Khaled O, Bashria AA, Maryam Nooman AlMallahi Y, Tan YC, Mahmoud M, Jaber H, Ramadan M (2022) An overview of smart irrigation systems using IoT. Energy Nexus 7: 100124. ISSN 2772-4271. https://doi.org/10.1016/j.nexus.2022.100124
3. Wang T, Xu X, Wang C, Li Z, Li D (2021) From smart farming towards unmanned farms: a new mode of agricultural production. Agriculture 11(2):145. https://doi.org/10.3390/agriculture11020145
4. Covaci and Gontean (2020) Piezoelectric energy harvesting solutions: a review. Sensors 20: 3512. https://doi.org/10.3390/s20123512
5. Saif H, Khan MB, Lee J, Lee K, Lee Y (2019) A high-voltage energy-harvesting interface for irregular kinetic energy harvesting in IoT systems with 1365% improvement using all-NMOS power switches and ultra-low quiescent current controller. Sensors 19(17):3685. https://doi.org/10.3390/s19173685
6. Shafiei N, Nasrollahzadeh M, Hegde G (2021) Chapter 10 – biopolymer-based (nano)materials for supercapacitor applications. In: Nasrollahzadeh M (ed) Biopolymer-based metal nanoparticle chemistry for sustainable applications, vol 2021. Elsevier, pp 609–671. https://doi.org/10.1016/B978-0-323-89970-3.00010-X. ISBN 9780323899703

7. Rasul G (2016) Managing the food, water, and energy nexus for achieving the sustainable development goals in South Asia. Environ Dev 18:14–25. https://doi.org/10.1016/j.envdev.2015.12.001
8. Sims B, Kienzle J (2017) Sustainable agricultural mechanization for smallholders: what is it and how can we implement it? Agriculture 7(6):50. https://doi.org/10.3390/agriculture7060050
9. Levidow L, Zaccaria D, Maia R, Vivas E, Todorovic M, Scardigno A (2014) Improving water-efficient irrigation: prospects and difficulties of innovative practices. Agric Water Manag 146: 84–94. https://doi.org/10.1016/j.agwat.2014.07.012
10. Canovas-Carrasco S, Garcia-Sanchez A, Garcia-Haro J (2018) On the nature of energy-feasible wireless nano sensor networks. Sensors 18:1356. https://doi.org/10.3390/S18051356
11. Phillips JD (2021) Energy harvesting in nanosystems: powering the next generation of the Internet of Things. Front Nanotechnol 3:5. https://doi.org/10.3389/fnano.2021.633931
12. Kurzweil (2015) Chapter 19 – electrochemical double-layer capacitors, moseley and garche. In: Electrochemical energy storage for renewable sources and grid balancing. Elsevier, pp 345–407. https://doi.org/10.1016/B978-0-444-62616-5.00019-X
13. Silva A, Liu M, Moghaddam M (2012) Power-management techniques for wireless sensor networks and similar low-power communication devices based on non-rechargeable batteries. J Comput Netw Commun 757291:10. https://doi.org/10.1155/2012/757291
14. Weselek A, Ehmann A, Zikeli S et al (2019) Agrophotovoltaic systems: applications, challenges, and opportunities. a review. Agron Sustain Dev 39:35. https://doi.org/10.1007/s13593-019-0581-3
15. Rai GD (2020) Non-conventional energy sources. Khanna Publishers, New Delhi
16. Jadhav S, Kalbande S (2012) Effect of raw material and moisture content on performance of piston press biomass briquetting machine. Green Farming 3(6):716–719
17. Pande M, Bhaskarwar AN (2012) Biomass conversion to energy. In: Baskar C, Baskar S, Dhillon R (eds) Biomass conversion. Springer, Berlin/Heidelberg. https://doi.org/10.1007/978-3-642-28418-2_1
18. Jadhav SK, Wakudkar H, Bhardwaj M, Soni R (2020) Effect of torrefaction on physio-chemical properties of paddy straw and its size reduction. Int J Curr Microbiol App Sci 9(01):7–18. https://doi.org/10.20546/ijcmas.2020.901.002
19. IEA (2022) Bioenergy. IEA, Paris. https://www.iea.org/reports/bioenergy. License: CC BY 4.0
20. Baruah D, Kalita P, Moholkar VS (2021) A comprehensive study on utilization of producer gas as IC engine fuel. In: Singh AP, Kumar D, Agarwal AK (eds) Alternative fuels and advanced combustion techniques as sustainable solutions for internal combustion engines. Energy, environment, and sustainability. Springer, Singapore. https://doi.org/10.1007/978-981-16-1513-9_6
21. IEA (2020) Outlook for biogas and biomethane: prospects for organic growth. IEA, Paris. https://www.iea.org/reports/outlook-for-biogas-and-biomethane-prospects-for-organic-growth. License: CC BY 4.0
22. Hotta SK, Sahoo N, Mohanty K (2019) Comparative assessment of a spark ignition engine fueled with gasoline and raw biogas. Renew Energy 134:1307–1319. https://doi.org/10.1016/j.renene.2018.09.049
23. Nijaguna BT (2021) Biogas technology. New Age International, New Delhi
24. Şen Z (2018) 4.12 Hydropower conversion. In: Dincer I (ed) Comprehensive energy systems. Elsevier, pp 545–572. https://doi.org/10.1016/B978-0-12-809597-3.00416-8
25. Kamran M (2021) Chapter 6 – hydro energy. In: Kamran M, Fazal MR (eds) Renewable energy conversion systems. Academic Press, pp 193–219. https://doi.org/10.1016/B978-0-12-823538-6.00007-5
26. Hepbasli A (2018) 4.4 Heat pumps. In: Dincer I (ed) Comprehensive energy systems. Elsevier, pp 98–124. https://doi.org/10.1016/B978-0-12-809597-3.00404-1
27. Rathore NS, Panwar NL (2021) Biomass production & efficient utilization for energy generation. New India Publishing Agency, New Delhi

28. Kalogirou SA (2014) Chapter 7 – industrial process heat, chemistry applications, and solar dryers. In: Kalogirou SA (ed) Solar energy engineering, 2nd edn. Academic Press, pp 397–429. https://doi.org/10.1016/B978-0-12-397270-5.00007-8
29. Ma Z, Venkataraman R, Farooque M (2009) Fuel cells – molten carbonate fuel cells | modeling. In: Garche J (ed) Encyclopedia of electrochemical power sources. Elsevier, pp 519–532. https://doi.org/10.1016/B978-044452745-5.00272-0
30. Kundu A, Jang JH (2009) Applications – portable | portable devices: fuel cells. In: Garche J (ed) Encyclopedia of electrochemical power sources. Elsevier, pp 39–45. https://doi.org/10.1016/B978-044452745-5.00349-X
31. Harb A (2011) Energy harvesting: state-of-the-art. Renew Energy 36(10):2641–2654
32. Yang Z, Zhou S, Zu Z, Inman D (2018) High-performance piezoelectric energy harvesters and their applications. Joule 2(4):642–697. https://doi.org/10.1016/j.joule.2018.03.011
33. Wu Y, Qiu J, Ji H, Zhou S (2018) Piezoelectric spring pendulum oscillator for animal/human motion energy harvesting. In: 2018 IEEE/ASME international conference on Advanced Intelligent Mechatronics (AIM), pp 774–779. https://doi.org/10.1109/AIM.2018.8452399
34. Nastro A, Pienazza N, Baù M, Aceti P, Rouvala M, Ardito R, Ferrari M et al (2022) Wearable ball-impact piezoelectric multi-converters for low-frequency energy harvesting from human motion. Sensors 22(3):772. https://doi.org/10.3390/s22030772
35. Li H, Lu J, Myjak MJ, Liss SA, Brown RS, Tian C, Deng ZD (2022) An implantable biomechanical energy harvester for animal monitoring devices. Nano Energy 98:107290
36. Catolico N, Ge S, John M (2016) Numerical modeling of a soil bore hole thermal energy storage system. Vadose Zone J 15. https://doi.org/10.2136/vzj2015.05.0078
37. Kubov VI, Dymytrov YY, Kubova RM (2016) Ltspice-model of thermoelectric peltier-seebeck element. In: IEEE 36th international conference on electronics and nanotechnology (ELNANO), pp 47–51. https://doi.org/10.1109/ELNANO.2016.7493007
38. Rees S, Adjali M, Zhou Z, Davies M, Thomas H (2000) Ground heat transfer effects on the thermal performance of earth-contact structures. Renew Sust Energ Rev 4:213–265. https://doi.org/10.1016/S1364-0321(99)00018-0
39. Akrouch G, Sánchez M, Briaud JL (2015) Effect of the unsaturated soil condition on the thermal efficiency of energy piles. Geotech Special Publication:1618–1627. https://doi.org/10.1061/9780784479087.146
40. Kovacic D (2019) Sensors and their application in precision agriculture. Graduate thesis, University of Osijek
41. Povh FP, Yasin M, Harun SW, Arof H (2014) Optical sensors applied in agricultural crops. In: Optical sensors- new developments and practical applications. InTech, New Delhi, pp 141–163
42. Paterova T, Prauzek M, Konecny J, Ozana S, Zmij P, Stankus M, Weise D, Pierer A (2021) Environment-monitoring IoT devices powered by a TEG which converts thermal flux between air and near-surface soil into electrical energy. Sensors 21(23):8098. https://doi.org/10.3390/s21238098
43. Aridi R, Faraj J, Ali S, Lemenand T, Khaled M (2021) Thermoelectric power generators: state-of-the-art, heat recovery method, and challenges. Electricity 2(3):359–386. https://doi.org/10.3390/electricity2030022
44. Zhu D (2011) Vibration energy harvesting: machinery vibration, human movement and flow induced vibration. In: Sustainable energy harvesting technologies – past, present and future. IntechOpen. https://doi.org/10.5772/25731
45. Zhai L, Gao L, Wang Z, Dai K, Wu S, Mu X (2022) An energy harvester coupled with a triboelectric mechanism and electrostatic mechanism for biomechanical energy harvesting. Nano 12:933. https://doi.org/10.3390/nano12060933
46. Lin Z, Chen J, Yang Y (2016) Recent progress in triboelectric nanogenerators as a renewable and sustainable power source. J Nanomater 5651613:24. https://doi.org/10.1155/2016/5651613
47. Rodriguez JC, Nico V, Punch JA (2019) Vibration energy harvester and power management solution for battery-free operation of wireless sensor nodes. Sensors 19(17):3776. https://doi.org/10.3390/s19173776

48. Rashika ME, Pushpalatha S (2018) A Wi-fi based animal health monitoring system. Int J Eng Res Technol 6(13):1–4
49. Roy S, Kabir MH, Salauddin M, Halim MA (2022) An electromagnetic wind energy harvester based on rotational magnet pole-pairs for autonomous IoT applications. Energies 15(15):5725. https://doi.org/10.3390/en15155725
50. Patel S, Novak N (2021) The pyroelectric energy harvesting and storage performance around the ferroelectric/antiferroelectric transition in PNZST. Journal of Materials Science 56(2):1133–1146
51. Thakre A, Kumar A, Song H-C, Jeong D-Y, Ryu J (2019) Pyroelectric energy conversion and its applications –flexible energy harvesters and sensors. Sensors 19(9):2170. https://doi.org/10.3390/s19092170
52. Safak M (2014) Wireless sensor and communication nodes with energy harvesting. J Commun Navig Sens Serv 1(1):47–66
53. Wang Y, Liu Y, Wang C, Li Z, Sheng X, Lee HG, Chang N, Yang H (2016) Storage-less and converter-less photovoltaic energy harvesting with maximum power point tracking for internet of things. IEEE Trans Comput Aid Desig Integr Circuits Syst 35(2):173–186
54. Kausar AZ, Reza AW, Saleh MU, Ramiah H (2014) Energizing wireless sensor networks by energy harvesting systems: scopes, challenges, and approaches. Renew Sust Energ Rev 38:973–989
55. Shaikh FK, Zeadally S (2016) Energy harvesting in wireless sensor networks: a comprehensive review. Renew Sust Energ Rev 55:1041–1054
56. Li J, Liu A, Shen G, Li L, Sun C, Zhao F (2015) Retro-vlc: enabling battery-free duplex visible light communication for mobile and IoT applications. In: Proceedings of the 16th international workshop on mobile computing systems and applications. ACM, pp 21–26. https://doi.org/10.1016/j.nanoen.2022.107290
57. Hawkes AM, Katko AR, Cummer SA (2013) A microwave metamaterial with integrated power harvesting functionality. Appl Phys Lett 103(16):163901
58. Shenck NS, Paradiso JA (2001) Energy scavenging with shoe mounted piezoelectrics. IEEE Micro 21(3):30–42
59. Mishra D, De S, Chowdhury KR (2015) Charging time characterization for wireless of energy transfer. IEEE Trans Circuits Syst II Exp Briefs 62(4):362–366
60. Tervo J, Manninen A, Ilola R, Hanninen H (2009) State-of-the-art of thermoelectric materials processing. Espoo, VTT Technical Research Centre of Finland 124:1–30
61. Ferrari M, Ferrari V, Marioli D, Taroni A (2006) Modeling, fabrication and performance measurements of a piezoelectric energy converter for power harvesting in autonomous microsystems. IEEE Trans Instrum Meas 55(6):2096–2101
62. Cuadras A, Gasulla M, Ferrari V (2010) Thermal energy harvesting through pyroelectricity. Sensors Actuators A Phys 158(1):132–139
63. Wang Z, Leonov V, Fiorini P, Van Hoof C (2009) Realization of a wearable miniaturized thermoelectric generator for human body applications. Sensors Actuators A Phys 156(1): 95–102
64. Leonov V, Torfs T, Fiorini P, Van Hoof C (2007) Thermoelectric converters of human warmth for self-powered wireless sensor nodes. IEEE Sensors J 7(5):650–657
65. Chen ZG, Han G, Yang L, Cheng L, Zou J (2012) Nanostructured thermoelectric materials: current research and future challenge. Prog Nat Sci Mater Int 22(6):535–549
66. Satyala N, Norouzzadeh P, Vashaee D (2014) Nano bulk thermoelectrics: concepts, techniques, and modeling. In: Nanoscale thermo-electrics. Springer, Cham, pp 141–183
67. Singh H, Lalchand CM (2012) Self-powered wearable health monitoring system. Adv Mater Res 403:3839–3846
68. Ramsay MJ, Clark WW (2001) Piezoelectric energy harvesting for bio-mems applications. In: Smart structures and materials: industrial and commercial applications of smart structures technologies, vol 4332. International Society for Optics and Photonics, pp 429–439

69. Zurbuchen A, Pfenniger A, Stahel A, Stoeck CT, Vandenberghe S, Koch VM, Vogel R (2013) Energy harvesting from the beating heart by a mass imbalance oscillation generator. Ann Biomed Eng 41(1):131–141
70. Discenzo FM, Chung D, Loparo KA (2006) Power scavenging enables maintenance-free wireless sensor nodes. In: Proceedings of the 6th international conference on complex systems
71. Sterken T, Fiorini P, Baert K, Puers R, Borghs G (2003) An electret-based electrostatic μ-generator in transducer, solid-state sensors, actuators and microsystems, In: 12th international conference proceeding transducers IEEE, vol. 3, issue 2, pp 1291–1294. https://doi.org/10.1109/SENSOR.2003.1217009
72. Sazonov E, Li H, Curry D, Pillay P (2009) Self-powered sensors for monitoring of highway bridges. IEEE Sensors J 9(11):1422–1429
73. Park KI, Xu S, Liu Y, Hwang GT, Kang SJ, Wang ZL, Lee KJ (2010) Piezoelectric $BaTiO_3$ thin film nanogenerator on plastic substrates. Nano Lett 10(12):4939–4943. https://doi.org/10.1021/nl102959k
74. Kwon J, Seung W, Sharma BK, Kim SW, Ahn JH (2012) A high performance PZT ribbon-based nanogenerator using graphene transparent electrodes. Energy Environ Sci 5(10): 8970–8975
75. Hwang GT, Park H, Lee JH, Oh S et al (2014) Self-powered cardiac pacemaker enabled by flexible single crystalline PMN-PT piezoelectric energy harvester. Adv Mater 26(28): 4880–4887. https://doi.org/10.1002/adma.201400562
76. Roundy SJ (2003) Energy scavenging for wireless sensor nodes with a focus on vibration to electricity conversion. Ph.D. dissertation, University of California
77. Starner T (1996) Human-powered wearable computing. IBM Syst J 35(3.4):618–629
78. Lai E, Redfern A, Wright P (2005) Vibration powered battery assisted passive RFID tag. In: Embedded and ubiquitous computing, pp 1058–1068
79. Enescu D (2019) Thermoelectric energy harvesting: basic principles and applications. In: Green energy advances. IntechOpen. https://doi.org/10.5772/intechopen.83495
80. Nesser H, Lubineau G (2021) Strain sensing by electrical capacitive variation: from stretchable materials to electronic interfaces. Adv Electron Mater 7(10):2100190. https://doi.org/10.1002/aelm.202100190
81. Petrović D, Jurišić M, Tadić V, Plaščak I, Barač Ž (2018) Different sensor systems for the application of variable rate technology in permanent crops. Tehnicki Glasnik 12(3):188–195. https://doi.org/10.31803/tg-20180213125928
82. Gill R, Chawla P (2022) Energy harvesting sensors based internet of things system for precision agriculture. In: 2nd International conference on innovative practices in technology and management (ICIPTM), pp 270–273. https://doi.org/10.1109/ICIPTM54933.2022.9754203
83. Oliveira LFP, Moreira AP, Silva MF (2021) Advances in agriculture robotics: a state-of-the-art review and challenges ahead. Robotics 10(2):52. https://doi.org/10.3390/robotics10020052
84. Tsouros DC, Bibi S, Sarigiannidis PG (2019) A review on UAV-based applications for precision agriculture. Information 10(11):349. https://doi.org/10.3390/info10110349
85. Chen H, Chen LJ, Albright TP (2007) Predicting the potential distribution of invasive exotic species using GIS and information-theoretic approaches: a case of ragweed (Ambrosia artemisiifolia L.) distribution in China. Chin Sci Bull 52(9):1223–1230. https://doi.org/10.1007/s11434-007-0192-2
86. Miskam MA, Nasiruddin AB, Rahim IA (2009) Preliminary design on the development of wireless sensor network for paddy rice cropping monitoring application in Malaysia. Eur J Sci Res 37(4):649–657

Challenges and Opportunities for Green Energy Harvesting in Sustainable IoT Systems

S. Sankaranath, M. Karthiga, S. Sountharrajan, and E. Suganya

1 Introduction

The Internet of Things (IoT) has been a rapidly growing trend in recent years, connecting an ever-increasing number of physical devices and appliances to the Internet. This interconnectivity has resulted in a vast amount of data being generated from these devices. IoT devices have a range of sensors, such as temperature, humidity, pressure, motion, and light, that collect data and then send it to the cloud for processing and analysis.

IoT devices are found in a variety of domains, including smart homes, transportation, healthcare, and industrial applications. In smart homes, IoT devices like smart thermostats, smart lighting, and security systems can be controlled through voice commands or mobile applications, offering enhanced automation, convenience, and energy efficiency. In transportation, IoT devices are being used to track and optimize the movement of vehicles, resulting in improved safety, reduced emissions, and more efficient logistics. In healthcare, IoT devices are enabling remote patient monitoring and personalized care, allowing patients to receive medical attention from the comfort of their homes. In industrial applications, IoT devices are being

S. Sankaranath (✉)
Department of EEE, Excel Engineering College, Namakkal, Tamilnadu, India

M. Karthiga
Department of CSE, Bannari Amman Institute of Technology, Sathyamangalam, Tamilnadu, India

S. Sountharrajan
Department of CSE, School of Computing, Amrita School of Engineering, Chennai, Tamilnadu, India

E. Suganya
Department of IT, SSN College of Engineering, Kalavakkam, Tamilnadu, India

A. Nella et al. (eds.), *Energy Harvesting Trends for Low Power Compact Electronic Devices*, EAI/Springer Innovations in Communication and Computing,
https://doi.org/10.1007/978-3-031-35965-1_7

used for predictive maintenance, production optimization, and supply chain management, among other use cases [1].

IoT devices are typically equipped with connectivity components such as Wi-Fi, Bluetooth, or cellular networks. This allows them to communicate with each other and with servers or cloud-based services, using various protocols like MQTT, CoAP, or HTTP. IoT applications use data analytics, machine learning, and artificial intelligence to process the vast amounts of data generated by IoT devices. These tools extract insights and actionable information that can help organizations to optimize their operations, improve decision-making, and create new business models [2].

However, IoT also presents significant challenges related to security, privacy, interoperability, and scalability. Security issues can arise due to the vulnerabilities of IoT devices, which can be targeted by hackers for malicious purposes. Privacy issues can also emerge from the collection and sharing of personal data by IoT devices. Interoperability issues can arise when devices from different manufacturers use different protocols, making it difficult for them to communicate with each other. Scalability issues can arise when many devices are connected, and the infrastructure and systems cannot handle the volume of data generated.

Addressing these issues requires a holistic approach that involves stakeholders from different domains and disciplines, including hardware and software developers, data scientists, policymakers, and end-users. Industry standards and regulations, as well as best practices for security and privacy, are being developed to address these challenges. As the number of connected devices continues to grow, it is essential to ensure that IoT is developed in a way that is secure, reliable, and privacy-preserving, while enabling innovation and growth in various domains. Wireless sensor networks (WSNs) are a type of Internet of Things (IoT) application that relies on a multitude of sensor nodes to perform distributed sensing and data processing tasks through wireless communication. These networks are composed of several key components, including sensor nodes, base stations, and a communication network.

Sensor nodes, which are the fundamental building blocks of WSNs, consist of a sensing unit, a processing unit, a memory unit, and a wireless communication unit. The sensing unit is responsible for detecting physical events or environmental parameters, such as temperature, humidity, light, sound, and motion. The processing unit performs local computations on the data collected by the sensing unit, while the memory unit stores both data and program instructions. The wireless communication unit enables the node to communicate with other nodes or base stations.

However, WSNs are not without their challenges, and one of their most significant weaknesses is their limited battery power. Since most sensor nodes rely on batteries with a finite lifespan, typically ranging from a few months to a few years, their operation faces multiple challenges due to the limited battery power. As shown in Fig. 1, a wireless sensor typically consists of the sensorial portion, the CPU, the transmitter, and a power source.

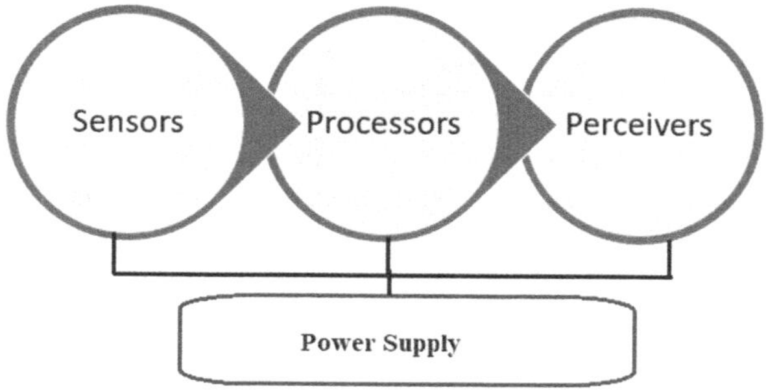

Fig. 1 Components of wireless sensor devices

2 Exigency for Energy Harvesting Techniques in IoT and WSN Environment

As the popularity of the Internet of Things (IoT) and wireless sensor networks (WSNs) continues to grow, the need for energy-efficient solutions becomes increasingly urgent. Energy harvesting techniques have emerged as a promising solution to address the power limitations of these devices [3]. These techniques enable IoT and WSN devices to extract energy from their surroundings and convert it into electrical energy to power themselves. There are several reasons why energy harvesting techniques are essential for IoT and WSN [4]:

- Firstly, these devices are often deployed in remote and inaccessible locations where battery replacement is challenging or impossible. Therefore, energy harvesting can ensure that the devices continue to operate without human intervention.
- Secondly, battery size and weight constraints limit the runtime of these devices, and energy harvesting techniques can extend their lifespan.
- In addition, energy harvesting techniques can reduce the carbon footprint associated with battery production and disposal.
- The use of rechargeable batteries and energy harvesting technologies reduces the need for frequent replacements and disposal, resulting in a lower environmental impact.

Various types of energy harvesting techniques can be used in IoT and WSN devices, including solar, thermal, mechanical, and electromagnetic energy harvesting [5]. These techniques can extract energy from different sources such as light, temperature differences, vibrations, and electromagnetic fields, respectively. Overall, energy harvesting techniques can significantly enhance the efficiency, reliability, and sustainability of IoT and WSN devices, making them more practical and cost-effective for diverse applications [6].

2.1 *Energy Harvesting Technique and Its Primary Architectures*

Energy harvesting is a process that involves extracting energy from the surrounding environment, such as light, heat, motion, or radio waves, and converting it into electrical energy that can power electronic devices. A typical energy harvesting device consists of a transducer that converts energy from the environment into electrical energy and a power management unit that regulates and stores the harvested energy [7]. There are four primary architectures of energy harvesting systems:

- Passive energy harvesting systems rely entirely on energy harvested from the environment and do not include any external power sources, such as batteries or power cables.
- Active energy harvesting systems, on the other hand, combine energy harvesting with energy storage and management, making them more reliable and robust.
- Hybrid energy harvesting systems use multiple energy harvesting techniques to increase efficiency and reliability.
- Finally, wearable energy harvesting systems are designed to harvest energy from human body motion, heat, or bio-signals to power wearable devices.

Energy harvesting systems offer a promising solution for powering electronic devices without relying on batteries or external power sources. Their primary architectures are specifically designed to optimize the harvesting of energy from various sources, making them practical and cost-effective for various applications [8]. The two different topologies for a wireless sensor node with energy harvesting components are shown in Figs. 2 and 3. Depending on how the system is configured to fit into one of the two architectures, its power generation component may include particular energy harvesting circuits that transform incoming light into direct current energy, units of power management that boost the effectiveness of storage components, and power generation that can store access from energy. Intense study is being done on all of these components, as well as the features and behaviors that are connected to them, with the goal of producing energy-autonomous wireless sensors that can be used in IoT environments [8].

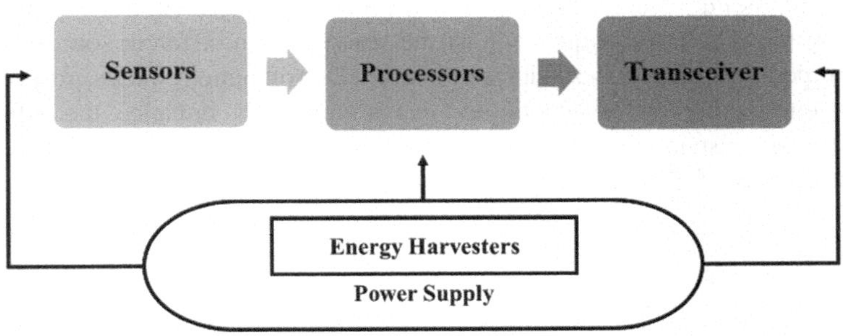

Fig. 2 Energy harvesters utilized wireless sensor architectures

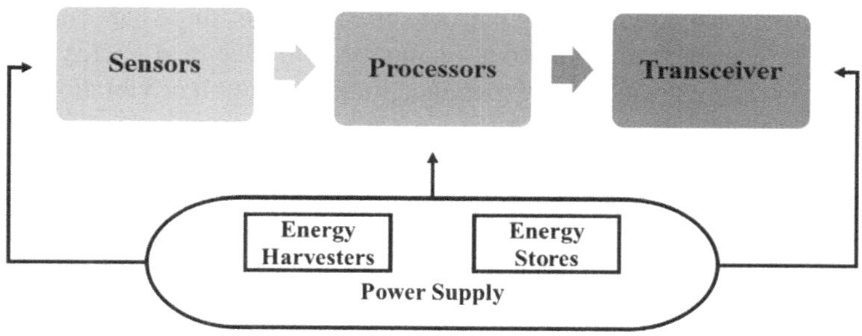

Fig. 3 Energy harvesters with stores-based wireless sensor architectures

3 Major Roots for Energy Harvesters

There are several sources available to harvest energy, including solar, thermal, kinetic, radio frequency (RF), magnetic, ambient light, and vibration energy. Solar energy can be converted into electrical energy using photovoltaic cells, while thermoelectric generators can convert temperature differences into electrical energy. Piezoelectric materials can convert mechanical energy into electrical energy from vibrations or motion, and RF harvesting can collect energy from wireless signals like Wi-Fi or cellular networks. Electromagnetic induction can convert magnetic energy from power lines or magnetic fields into electrical energy. Photovoltaic cells can also convert light from indoor lighting or other ambient sources into electrical energy, while electromagnetic induction, piezoelectric materials, and electrostatic generators can convert mechanical vibration or motion into electrical energy. Overall, energy harvesting provides a sustainable power source that can be obtained from a variety of sources for electronic devices. When using one energy source alone is insufficient, it may be necessary to combine many different energy sources to produce the necessary power for electronic equipment [9].

4 Ambient Energy Sources

Ambient energy harvesters are designed to convert energy from the environment into electrical energy for powering electronic devices. These devices come in different types, including:

- *Solar energy harvesters*: Utilizing photovoltaic cells, these harvesters convert solar energy into electricity, providing power for low-power electronic devices like sensors and wireless nodes.
- *Thermal energy harvesters*: These harvesters convert temperature differences into electricity using thermoelectric generators. They are ideal for powering devices that generate heat, such as combustion engines or industrial processes.

- *Kinetic energy harvesters*: Using piezoelectric materials, these harvesters convert mechanical energy from vibrations or motion into electrical energy. They are ideal for powering devices that experience frequent motion or vibrations, such as vehicles or machinery.
- *Radio frequency (RF) energy harvesters*: These harvesters scavenge energy from wireless signals like Wi-Fi or cellular networks to power electronic devices, particularly wireless devices like sensors or wearables.
- *Magnetic energy harvesters*: By utilizing electromagnetic induction, these harvesters convert magnetic energy from power lines or magnetic fields into electrical energy. They are best suited for devices that operate near power lines or generate their magnetic fields, like electric motors.
- *Ambient light energy harvesters*: These harvesters convert light from ambient sources, including indoor lighting, into electrical energy using photovoltaic cells. They are ideal for powering devices that operate indoors or in low-light environments.
- *Vibration energy harvesters*: These harvesters convert mechanical vibration or motion into electrical energy using piezoelectric materials, electromagnetic induction, or electrostatic generators. They are ideal for powering devices that experience frequent motion or vibrations, such as vehicles or machinery.

Overall, ambient energy harvesters provide a sustainable and reliable power source for electronic devices, and the different types can be optimized for specific environmental conditions and device requirements.

4.1 Solar Power Harvesters

Solar energy harvesters convert solar energy into electrical energy using photovoltaic cells [10]. They offer several advantages, including being a renewable and abundant source of energy, requiring low maintenance, and being environmentally friendly. However, they are dependent on direct sunlight and can have high initial costs. Some real-world applications of solar energy harvesters include solar-powered streetlights, water pumps, and remote sensors for environmental monitoring. These implementations demonstrate the practicality and effectiveness of solar energy harvesting in various industries [11]. In conclusion, solar energy harvesting is a promising technology for providing sustainable and renewable power to electronic devices. Its implementation in real-world applications highlights its potential in addressing energy needs in remote and off-grid areas. The most popular construction, as shown in Fig. 4, contains storage components to guarantee continuous operation of the device powered by solar energy collecting components (during times when ambient light is not accessible, such as at night and when clouds are present). Super-capacitors, rechargeable batteries, or a mix of the two can all be used as energy storage, each with their own benefits and drawbacks. When the amount of power generated exceeds the device's consumption, when it is sleeping or inactive,

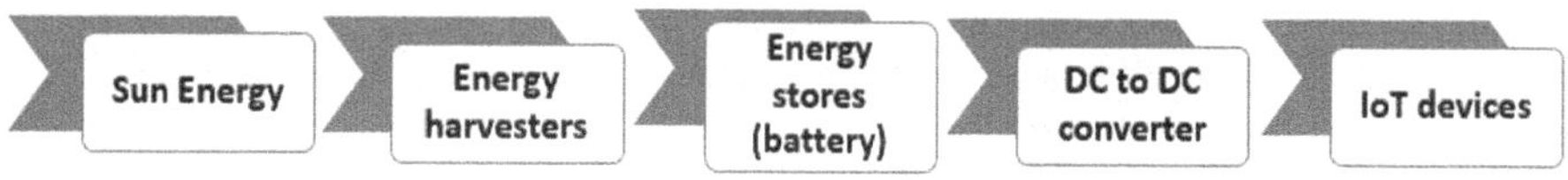

Fig. 4 Needed components for seizing solar energy

or both, the excess power is saved, and it is used when the energy source is absent or when the system calls for it [12].

4.2 Thermal and Kinetic Energy Harvesters

Thermal energy harvesters utilize thermoelectric generators to convert temperature differences into electrical energy, whereas kinetic energy harvesters employ piezoelectric materials to convert mechanical energy from motion or vibrations into electrical energy. Both have their own set of advantages and disadvantages, as well as real-time implementation examples.

Advantages of Thermal Energy Harvesting

- No fuel required: Thermal energy harvesting is an eco-friendly and cost-effective power source that does not require any fuel.
- Stable power supply: Thermal energy harvesting provides a steady power supply, even in harsh environmental conditions, unlike solar or wind power.
- High efficiency: Thermoelectric generators can convert up to 10% of the thermal energy into electrical energy, making them highly efficient.

Disadvantages of Thermal Energy Harvesting

- Limited temperature range: Thermal energy harvesting is restricted by the temperature difference between the hot and cold sides of the generator, making it less effective in applications with small temperature differences.
- Cost: The cost of thermoelectric generators can be high, which may restrict their usage in some applications.
- Complexity: Thermal energy harvesting systems can be complex, requiring careful design and optimization to achieve high efficiency.

Real-Time Implementation Examples of Thermal Energy Harvesting

- Waste heat recovery: Thermal energy harvesting can recover waste heat from industrial processes and convert it into usable electrical energy.
- Automotive exhaust heat recovery: Thermoelectric generators can recover heat from automotive exhaust systems and use it to power electronic components.
- Remote sensors: Thermal energy harvesting can power remote sensors used in environmental monitoring or building automation systems.

Advantages of Kinetic Energy Harvesting

- Abundant source: Kinetic energy is present in many environments, making it a potentially abundant source of power.
- Low maintenance: Kinetic energy harvesters have no moving parts and require little to no maintenance.
- Durability: Piezoelectric materials are highly durable and can withstand harsh environmental conditions.

Disadvantages of Kinetic Energy Harvesting

- Limited power output: Kinetic energy harvesting typically generates low levels of power, making it unsuitable for high-power applications.
- Limited frequency range: Piezoelectric materials have a restricted frequency range, making them less effective in applications with high-frequency vibrations.
- Environmental variability: Kinetic energy sources can vary significantly in intensity and frequency, making it challenging to design an effective energy harvesting system.

Real-Time Implementation Examples of Kinetic Energy Harvesting

- Wireless sensors: Kinetic energy harvesters can power wireless sensors used in environmental monitoring or industrial automation.
- Wearable electronics: Kinetic energy harvesters can power wearable electronics, such as fitness trackers or smartwatches.
- Automotive vibration energy harvesting: Kinetic energy harvesting can recover energy from automotive vibrations and use it to power electronic components.

In summary, both thermal and kinetic energy harvesting provide sustainable and reliable power sources for electronic devices, with various implementation examples in real-world applications.

4.3 RF Energy Harvesters

According to Fig. 5, sources of RF or electromagnetic signals in the frequency range of 3 kHz to 300 GHz include radio and satellite stations, wireless Internet, and broadcasts for digital multimedia. These signals can be transformed into electrical energy using an antenna and a rectifier circuit [4]. There are various degrees of this energy everywhere around us (both within and outside), and it is always present (day or night). When it comes to low-energy Internet of Things devices, radio frequency waves energy harvesting is the best option in numerous situations [13].

IoT devices have the ability to power RF energy harvesters (RFEHs) for monitoring the environment and human health.

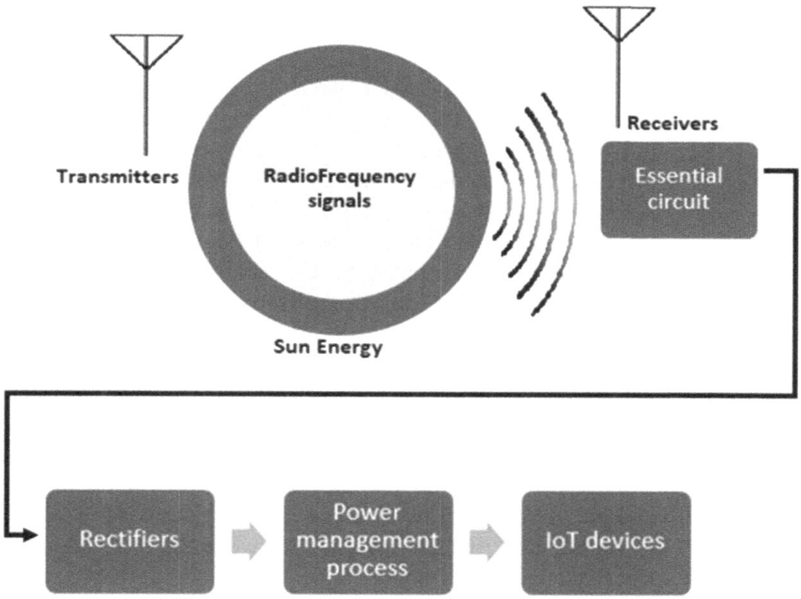

Fig. 5 Conversion of sun energy to power for IoT circuits

5 Mechanical Energy Sources

In IoT environments, mechanical energy harvesters are utilized to convert mechanical energy into electrical energy, which eliminates the need for external power sources or batteries to power electronic devices and sensors. Vibration, motion, and pressure harvesters are common types of mechanical energy harvesters used in IoT environments. Vibration harvesters use piezoelectric or electromagnetic materials to generate electricity from ambient vibrations for applications like environmental monitoring, industrial automation, and structural health monitoring. Motion harvesters use piezoelectric, electromagnetic, or electrostatic materials to generate electricity from motion and can power wearable devices and industrial automation sensors [14]. Pressure harvesters generate electricity from pressure or strain and can be used to power smart flooring or other low-power IoT devices. The benefits of mechanical energy harvesters in IoT environments include their low maintenance requirements, long lifespan, and eco-friendliness [15]. However, they have some limitations, such as limited power output and sensitivity to environmental factors. Hence, precise design and optimization are crucial for achieving optimal performance in specific applications.

6 Human Energy Harvesters

Human energy harvesters are devices that convert the energy generated by the human body into electrical energy. They have shown great potential for powering Internet of Things (IoT) devices, especially in wearable electronics and healthcare applications. However, several challenges need to be addressed to ensure their effective use in IoT environments. One example of a human energy harvester is the thermoelectric generator (TEG), which can convert body heat into electricity. However, the efficiency of TEGs is limited by the small temperature difference between the human body and the environment. Researchers are exploring new materials with higher thermoelectric properties to overcome this challenge. Another example is kinetic energy harvesters, which can convert human motion into electricity. However, their output power is low, making them only suitable for low-power applications. Researchers are exploring new materials and designs to enhance their efficiency and increase their power output. Piezoelectric materials are also used as human energy harvesters, but their sensitivity to temperature and humidity can affect their performance in different environments [16]. Researchers are developing more stable and reliable materials to overcome this challenge. In addition to technical challenges, the integration of human energy harvesters with other electronic components and the need for efficient power management systems are also important considerations. User comfort, safety, and privacy are also critical, especially in healthcare applications. Thus, human energy harvesters have promising potential for powering IoT devices, but further research and development are necessary to overcome current challenges and ensure their effective use in practical applications.

7 Biomechanical Energy Harvesters

Biomechanical energy harvesters are devices that generate electrical energy from the movement of human or animal bodies. They hold great promise for powering IoT devices in a variety of applications, including healthcare, sports, and environmental monitoring. However, there are several challenges that must be overcome to optimize their performance and reliability in IoT environments. Here are some examples of biomechanical energy harvesters and their associated challenges [17]:

- *Implantable energy harvesters*: These harvesters generate electricity from the movement of internal organs or tissues and can power medical implants like pacemakers or artificial organs without external batteries or wires. However, integrating them with the human body requires careful consideration of biocompatibility, safety, and long-term stability.
- *Biofuel cells*: These harvesters generate electricity from the chemical energy of biological fluids like blood or sweat and have potential in wearable devices like smartwatches or fitness trackers. However, their efficiency and stability are

impacted by the variability of biological fluids, pH, and temperature and the need for biocompatible materials and enzymes.

- *Muscle-powered generators*: These harvesters generate electricity from muscle contractions, such as those in the arm or leg, and have potential in IoT devices in sports or military applications. However, their output power and efficiency depend on the user's strength and endurance, and their integration with the body requires careful consideration of ergonomics and safety.

The challenges of biomechanical energy harvesters in IoT environments also include the need for efficient energy storage and management systems, integration with wireless communication protocols, and reliable performance under different environmental conditions. Thus, biomechanical energy harvesters hold great promise for powering IoT devices in various applications. However, addressing the current challenges requires further research and development to ensure their effective use in practical applications.

8 Bioenergy Sources

Bioenergy harvesters refer to devices that produce electrical energy from biological sources, such as microorganisms, plants, and biomass waste. They offer a sustainable and renewable energy solution for a range of applications, such as biofuel production, electricity generation, and wastewater treatment. However, for them to be practical and scalable, a number of factors must be optimized and integrated into functional systems. For example, microbial fuel cells (MFCs) convert the metabolic activity of microorganisms found in wastewater or soil into electricity. MFCs could be an inexpensive and sustainable energy source for remote sensing or wastewater treatment. Nevertheless, the power output and stability of these cells rely on a range of factors, including the electrode material, the microbial community's diversity and activity, and the design of the system.

Photosynthetic bioelectrochemical systems (BESs) [18] generate electricity from the photosynthesis of algae or plants coupled with the electrochemical activity of fungi or bacteria. These systems offer carbon-neutral and sustainable energy for food production, carbon capture, and biomass production. However, their stability and efficiency are subject to various factors, such as the light intensity, the photosynthetic activity, and the microbial activity and diversity. Bioelectrochemical systems convert organic waste or biomass, like agricultural waste and food waste, into hydrogen gas or electricity. These systems can offer decentralized and sustainable energy for off-grid areas and rural communities. The efficiency and scalability of these systems depend on the quality of the feedstock, microbial activity, and the design and operation of the system. In addition, the challenges of bioenergy harvesters include the need for efficient energy storage and conversion systems, the integration with existing energy infrastructures, and social and regulatory acceptance.

9 Hybrid Energy Harvesters

Hybrid energy harvesters, which harness multiple energy sources, offer a promising solution for powering Internet of Things (IoT) devices. By combining solar, thermal, and mechanical sources, hybrid harvesters can provide a steady and reliable energy supply for various IoT applications, overcoming the limitations of individual sources. However, there are challenges that need to be addressed to optimize their performance and integration in IoT environments. Several examples of hybrid energy harvesters and their challenges include:

- Solar-thermoelectric generators (STEGs): STEGs can generate electricity from both solar radiation and temperature gradients. While they have potential for low-cost energy in smart cities or homes, their efficiency and reliability depend on material optimization, heat transfer mechanisms, and system design.
- Triboelectric-piezoelectric generators: These generators can produce electricity from mechanical energy, like footfalls or wind, making them a good candidate for wearable or remote sensing applications. Challenges include optimizing material properties, interface design, and integrating with wireless communication protocols.
- Thermoelectric-electromagnetic generators: These generators can produce electricity from both temperature gradients and mechanical vibrations, making them useful in transportation or industrial applications. Challenges include optimizing thermoelectric and electromagnetic materials, heat and vibration transfer mechanisms, and system design.

Challenges for hybrid energy harvesters in IoT environments also include the need for efficient energy management and storage systems, wireless communication protocols, and data processing. Additionally, durability and reliability must be optimized for various environmental conditions [19]. The various energy sources utilized in IoT-based wireless sensor environment for energy harvesting are discussed in Table 1.

10 Energy Harvesting Models vs Consumption Models

In the world of IoT, the use of energy harvesting and consumption models is crucial to ensure sustainable and reliable power supply to the devices. Energy harvesting models are employed to convert various ambient energy sources into electrical power, whereas consumption models are utilized to manage and optimize power usage. Here are some of the energy harvesting and consumption models that are widely used in IoT.

Table 1 Various energy sources for energy harvesting

S. no	Energy type	Source
1	Light energy	Energy from sunlight
2	Kinetic energy	Energy obtained by vibration or rotation or motion of objects
3	Thermal energy	Industrial sources or human sources
4	Atmospheric energy	Energy obtained by pressure or gravitational force
5	Water energy	Energy obtained from waves or water resources
6	Chemical energy	Energy obtained due to diffusion or nuclear fission

10.1 Energy Harvesting Models

- Solar energy harvesting: Solar cells are used to transform solar radiation into electrical power, and this model is commonly used in outdoor IoT applications, such as environmental monitoring or smart agriculture.
- Thermal energy harvesting: Thermoelectric modules or thermoelectric generators are used to convert temperature gradients into electrical power, and this model is often used in indoor IoT applications, such as smart homes or buildings.
- Mechanical energy harvesting: Piezoelectric or electromagnetic generators are used to convert mechanical movements or vibrations into electrical power. This model is commonly used in wearable or remote IoT applications, such as health monitoring or tracking.
- RF energy harvesting: RF harvesting technology is used to transform radio frequency signals into electrical power, and this model is usually used in wireless IoT applications, such as communication devices or wireless sensors.

10.2 Consumption Models

- Duty cycling: This model involves turning devices on and off at specific intervals to conserve energy, and it is commonly used in low-power IoT applications, such as environmental monitoring or sensor networks.
- Power management: Power consumption is managed by regulating the operating voltage or frequency of the device, and this model is often used in battery-powered IoT devices, such as smart meters or wearable devices.
- Energy harvesting-aware scheduling: Power consumption is scheduled based on the available energy harvested by the device, and this model is commonly used in hybrid energy harvesting systems, such as thermal and solar or mechanical and thermal systems.
- Energy storage: Energy is stored in a battery or capacitor and used when the harvested energy is insufficient, and this model is commonly used in energy harvesting systems that have intermittent or unpredictable energy sources.

Table 2 Potential power production using various energy harvesters

S. no	Energy harvesters	Power production
1	Thermoelectric	10 μW to 1 kilowatt
2	Photovoltaic	1 μW to 1 megawatt
3	Piezoelectric	10 μW to 100 W
4	Electrodynamic	0.1 μW to 1 MW
5	Pyroelectric	0.1 μW to 1 mW
6	Magnetic	10 μW to 1 MW
7	Bioenergy	0.1 μW to 1 mW

Overall, energy harvesting and consumption models are critical to ensuring that IoT devices have reliable and sustainable power supplies. By optimizing these models, IoT devices can operate effectively and efficiently in various environments and applications [20]. The obtained power using various energy harvesters for achieving sustainability in IoT-based wireless sensor domain is illustrated in Table 2.

11 Edge IoT and IoT Devices

The Internet of Things, or IoT for short, creates ambient intelligence, which relies on connectivity of the people and makes use of the electronic environment's adaptability and sensitivity to meet needs in the immediate context. IoT-based system deployment in remote places necessitates long periods of unattended operation. Constant power supply and energy efficiency are the main obstacles to accomplishing this goal. Energy harvesting enables on-site charging of the storage devices, which results in continuous operation of the sensor nodes. Edge devices also manage the information flow among boundaries of two networks. In essence, service providers use edge devices, which serve as network entrance and exit points. Processing, filtering, translating, storing, transmitting, and routing data throughout the network are some of their main responsibilities. IoT-based apps utilize edge devices to achieve advanced intelligence systems with features and to acquire more processing power as shown in Fig. 6. Edge computing is the use of logical physical locations and decentralized operations.

Energy is obtained from outside sources and turned into electrical energy using an energy collecting equipment in the process known as energy harvesting. The transducer, also known as an energy harvesting unit, collects energy from nearby sources and transforms it into electrical energy. The goal of the interface unit is to get the most energy possible out of the harvesting unit while adjusting the energy level to be compatible with the particular receiver or load. Various power management techniques, such as voltage control or rectification, have been used to achieve this. Different sensors, transducers, and other electronic circuits could be used in the receiver. The components of energy gathering system are illustrated in Fig. 7. There is a lot of energy in the surroundings that can be transformed into electricity and used

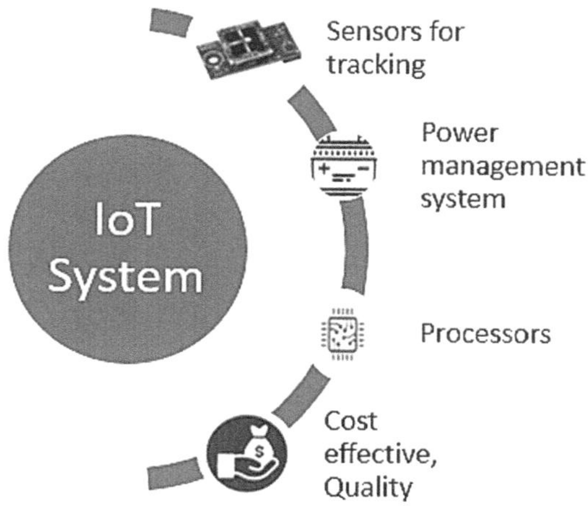

Fig. 6 Overview of Internet of Things system

to power various circuitry. In the same logic, it enables novel uses in distant, undersea, and similar tough regions where batteries and traditional power are impractical. Energy harvesting is also relatively low-maintenance and ecologically benign.

11.1 IoT Energy Harvesting Requirements

In particular when IoT are situated in difficult-to-reach locations and routine battery maintenance is not possible, energy harvesting is a promising alternative to power IoT. The device's lifespan is increased by using an energy harvesting strategy, which also gets rid of the power source from fixed charge batteries. The subsequent points highlight the significance of energy harvesting technology in Internet of Things (IoT) applications.

12 Constraints and Potentials of Energy Collecting Technology

Limitations and prospects of energy collecting technology are as follows:

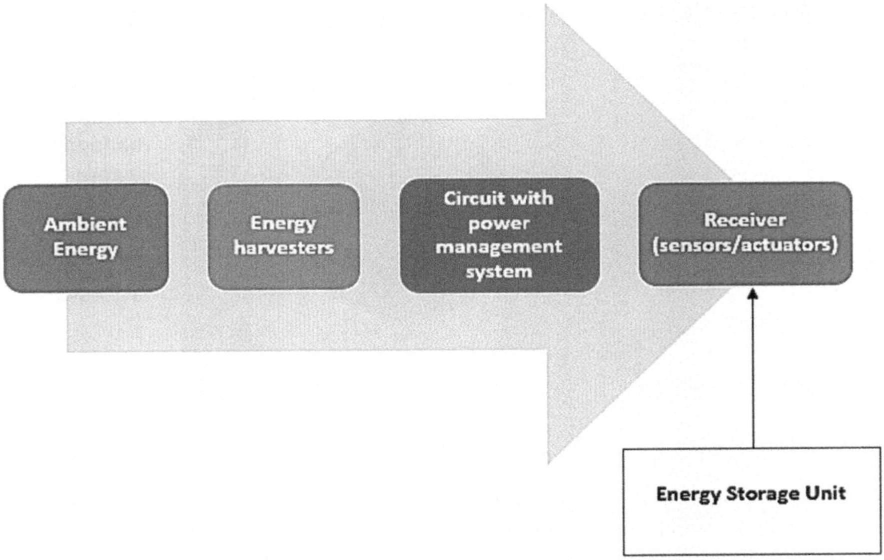

Fig. 7 Components in an energy gathering model

12.1 Constraints

The following are the barriers to the widespread use of energy harvesting techniques in the IoT domain:

- The most significant restriction to take into account is the high cost in comparison to traditional batteries.
- The size of the harvesting module, which grows in accordance with energy demands, is the second hurdle.
- An energy storage unit is necessary for the continuous operation of sensor nodes.

12.2 Potentials

Energy harvesting technology has a strong chance of becoming widely used, even though the added cost prevents the mass production of harvesting modules and some technologies' power requirements might not be compatible with harvesting technology. Here is a short list of a few of them:

1. An energy harvesting module's projected lifespan is greater than 5 years. According to a survey, certain energy collecting devices have been operating without hardware deterioration for the past 15 years. As a result, inaccessible regions do not require routine hardware maintenance.

2. Further reducing consumption of energy and improving the viability of energy harvesting is the development of advanced low-power electronic components in conjunction with cloud computing for data processing.
3. Building automation, where energy harvesting techniques can lower the cost of copper cables, materials, installation, and maintenance, is a significant area for energy savings.
4. As the IoT business expands, more energy harvesting modules will be used, which will lower the price.

13 Energy Harvesting Requirements

An essential requirement for the effective deployment of energy harvesting techniques in various applications is interoperability across various end systems. The three common policies represented in figure are utilized to meet the energy requirement goal. The EnOcean Alliance conforms to communication profile standardization, specifically the EnOcean Equipment Profiles, to enable seamless communication among its entire product range, which includes wireless switches, rechargeable sensors, and controls. EnOcean wireless standard is designed for energy harvesting sensor networks as well as wireless sensors and its networks with ultralow power consumption. The energy collection platform is depicted in Fig. 8 wherein the energy harvesters, energy stores, interfacing module, and low-power electrical module comprise the hardware portion.

Then the suitable backend software is implemented to create a complete energy gathering application. Figure 9 details about the market share for energy harvesting schemes followed by different industries. Among the various industrial sectors, energy harvesting schemes are mainly followed in residential and commercial places.

14 Applications in Socioeconomics

The primary driver of the market expansion for energy harvesting is the replacement of battery-operated power sensors with rechargeable wireless sensors based on energy harvesting. Particularly in offshore oil and gas systems and sensitive military sectors, sensors for rechargers are becoming the primary option for deployment in distant or inaccessible sites, and it is not viable to change the batteries often. Additionally, since they don't contain any chemicals or metals that are damaging to the environment, autonomous or rechargeable sensors are environmentally beneficial.

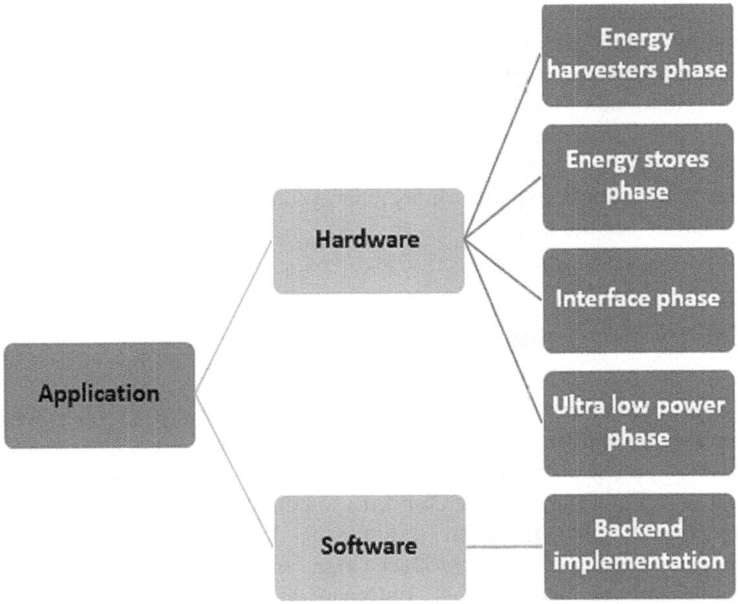

Fig. 8 Flow of energy collection platform

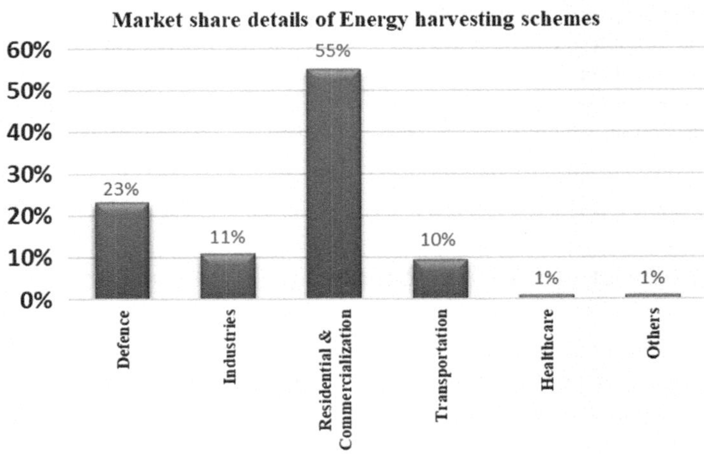

Fig. 9 Energy harvesting market share details for different industries [21]

14.1 Advantages for Basic Industries

The primary industry sector's energy harvesting market in the year of 2014 had a size of EUR 130 million, and it was expected to grow by the end of 2020 to EUR 380. In the primary industry sector, energy harvesting devices are used to improve the availability and dependability of industrial processes.

14.2 Advantages for the Defense

Energy harvesting technology is being used more frequently in the defense industry as it paves the way for the replacement of bulky battery packs with autonomous or rechargeable devices. Drones, autonomous airborne unmanned vehicles powered by solar photovoltaic energy, are a tremendous advancement for military use. The most recent analysis estimates that by 2020, defense applications based on energy harvesting will cost EUR 730 million.

14.3 Benefits to the Transportation Sector

Different modes of transportation offer a variety of energy sources that can be captured in various ways. For instance, solar energy is collected by solar panels placed on the road and transformed into electricity by people walking or driving. Similar to this, vibration energy produced by various modes of transportation can be used for energy harvesting. By 2020, the transportation industry is predicted to reach its aim of EUR 310 million.

14.4 Benefits of Residential and Business Properties

The future trend of home automation and smart housing will depend heavily on energy harvesting. This is due to the fact that energy harvesting technology will result in smaller mobile device batteries. Energy harvesting allows for rechargeable power supplies with minimal installation and maintenance costs, which enables the rapid expansion of the Internet of Things (IoT). At the end of 2020, the market value of this industry is anticipated to be EUR 1750 million. IoT devices have interacted with the network using a variety of communication protocols, each of which has advantages and disadvantages. Table 3 includes various communication protocols and their individual qualities.

Table 3 Merits and demerits of energy harvesters

Energy harvesters	Merits	Demerits
Solar	Simple, inexpensive, and effectual	Requires overcharge protection, unpredictable due to climatic changes
Mechanical	Protected, consistent, foreseeable output, higher energy output	Limit lifespan, degrade over time, issue in size, inadequate practical operative duration
Thermal	Long lifespan, no movable parts, suitable for body-worn applications	Low energy, higher cost, lower power output
Biochemical	Accessibility of constant sources of fuel	Sensitive to external atmosphere, no commercial, limited life, dependability of organic components

15 IoT Energy Sources' Features

- *Scalability*: The power supply for Internet of Things devices needs to be scalable. The devices must be positioned in a variety of locations to gather data and connect with the gateways in order to support the widespread deployment of IoT services and applications. Because the devices might be in difficult-to-reach places, they must operate on their own without human assistance. Batteries are not a feasible alternative because they likewise require little maintenance.
- *Maintenance-free*: The majority of Internet of Things applications require a maintenance-free energy supply. The majority of IoT use cases require several devices. It is not practical to wire these devices to the primary power source because doing so would limit their mobility and raise the overall deployment cost. Batteries are also not viable since frequent battery maintenance is problematic given the quantity of devices, cost of batteries, and scope of repair costs.
- *Mobility assistance*: Many IoT applications require energy sources that facilitate mobility. IoT devices are frequently on the move, so it is important to design the devices and power management systems with this in mind.
- *Long life*: Many IoT applications call for energy sources with a long-life span and low maintenance requirements. In some Internet of Things (IoT) applications, such as structural health monitoring, embedded wireless sensors must be installed inside of buildings, bridges, etc., and they must last for many years. To extend the lifespan of the device in various applications, it is necessary to improve both the energy storage and wireless communication.
- *Flexibility*: Given the wide range of IoT services and applications, IoT energy sources must be adaptable in terms of size and capacity. For instance, in applications for health monitoring, a tiny gadget will be implanted inside the body of the user to sense important data and transmit it to a personal device. Long-lasting little batteries or batteries with energy harvesting capabilities are needed for this. Due to their size and placement, many other applications, like parking meters, may be connected to the main power supply or may profit from huge photovoltaic cells.

- *Low cost*: Installing a lot of devices is necessary for many IoT applications. If these gadgets and, thus, their power sources are expensive, the application will either generate little money or be prohibitively expensive, which will restrict its appeal.
- *Sustainability*: The Internet of Things will comprise billions of devices at various scales, and because every one of them uses power, it will have an impact on the environment. To prevent the depletion of natural resources and preserve ecological balance, IoT power solutions must be sustainable. This highlights the significance of energy harvesting power sources for IoT as alternatives to traditional, unsustainable power sources.
- *Environment-friendly*: The proliferation of IoT services will have a negative influence on the environment because of the vast amount of batteries that are thrown away, for example, more than 125,000 tons of batteries are thrown away annually in the USA. The issue would get worse as used AA batteries would complete six orbits around the planet. It is very important to address this issue, and energy harvesting methods provide numerous potential answers.

16 Energy Harvesting Merits and Demerits

Due to the development of ultralow-power multipoint control units, energy harvesting did not pose any problems for embedded developers (MCUs). But as portable device geometries grew smaller and battery technology lagged behind, energy harvesting became a serious topic of discussion. Photovoltaic cells possess the greatest power density and output among all energy sources. The various merits and demerits of energy harvesters in the IoT domain are discussed in Table 3.

The technologies and gadgets utilized worldwide for energy harvesting are depicted in Table 4.

IoT applications are power intensive in nature due to continuous sensing, transmission, and perseverance. The best energy harvesters for various IoT applications according to their power density strategy are discussed in Table 5.

According to IDC, there could be up to 41 billion Internet-connected systems deployed globally by 2025. Among the most difficult practical problems of an IoT system once implemented is its short battery duration. As numerous devices in terms of million range are deployed globally in IoT environment, it is practically impossible to replace the batteries manually. Furthermore, device connectivity may get lost at any time due to down power range of these batteries and so the total invested cost is useless. To tackle this issue, battery life could be maximized by introducing more efficient power devices. For this the new green energy harvesting techniques could be utilized to tackle the power issues. Secondly, low-power designs like low-power ICs and power-concentrated communication protocols could be utilized in IoT systems. This option will optimize the necessity of when and how much a communication should be taken place in an IoT system. The technological advancements with respect to energy harvesting in the IoT domain are illustrated in Fig. 10 which shows there is a tremendous increase in technology with respect to saving energy.

Table 4 Communication protocols with necessary qualities

Communication protocol	Power	Data rate	Distance covered
Wi-Fi	50 mW	250 kb/s – medium rate applications	25 m
ZigBee	>500 mW	>100Mbs – high-rate applications	10 m
LoRa	100 mW	1Kbs – low-rate applications	500 m
Bluetooth	10 mW	1Mbs – high-rate applications	10 m

Table 5 Consolidation information of energy harvesting models

Energy source	Harvesters	Power density	Suitable for
Solar	Solar cells	~100 mW	Calculators, wireless sensor networks, indoor and outdoor monitoring
Mechanical	Electrostatic, electromagnetic, piezoelectric	430 μW, 493 μW, 963 μW	IoT systems, GPS trackers, hearing aids, pacemakers, timers ICs
Thermal	Thermoelectric generators	~1 W	Wearables
RF signals	RF energy harvesters	>1 mW	Wearables, IoT sensors, GPS trackers, communication systems, Bluetooth and RF tags, backscatter communication systems

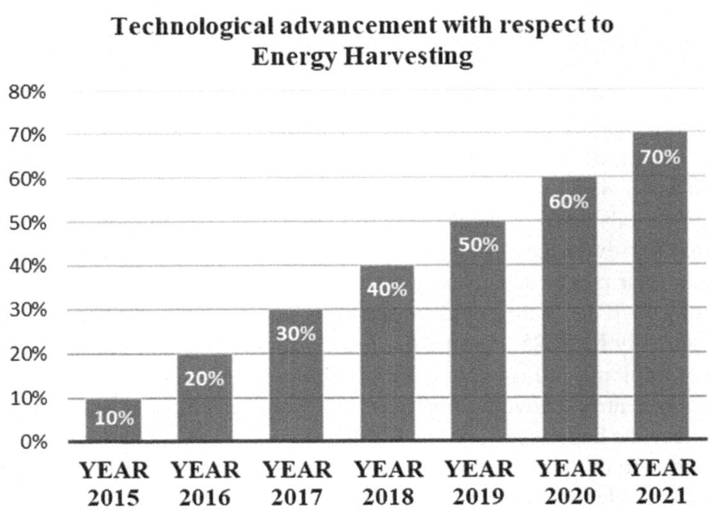

Fig. 10 Technological advancements with respect to the energy harvesting

17 Conclusion

Enhancing the dependability and lifetime of IoT systems requires energy harvesting. Energy sources from which it is intended to be gathered energy are not always available, which is one disadvantage of energy harvesting systems. This analysis of the energy collecting sources explains the efforts that have been done and novel structures that have been developed to overcome these limitations. IoT products have aided the aerospace industry, military technology, gas sensing, environmental monitoring, and patient healthcare systems. The Internet of Things also has potential in wearable technology. Different levels of technological sophistication are possible because of a variation of ambient energy sources and reliable energy gathering techniques. This series is suitable for different types of IoT applications and devices. In the prior literature review, a number of practical energy harvester implementations that supply power to drive Internet of Things devices are surveyed. Energy harvesting technology, however, is not yet widely used worldwide due to a number of technological constraints. To meet the potential demands of the world, these issues must be fixed.

References

1. Kiziroglou ME, Yeatman EM (2012) Materials and techniques for energy harvesting. Woodhead Publishing, London
2. Defeo C (2015) Energy harvesting and the Internet of Things. Elsevier, Amsterdam
3. Mukherjee S, Biswas GP (2018) Networking for IoT and applications using existing communication technology. Egypt Inform J 19(2):107–127
4. Sanislav T, Mois GD, Zeadally S, Folea SC (2021) Energy harvesting techniques for Internet of Things (IoT). IEEE Access 9:39530–39549. https://doi.org/10.1109/ACCESS.2021.3064066
5. Garg N, Garg R (2018) Energy harvesting in IoT devices: a survey. In: Proceedings of the international conference on intelligent sustainable systems, ICISS 2017
6. Sanislav T, Zeadally S, Mois GD, Folea SC (2018) Wireless energy harvesting: empirical results and practical considerations for Internet of Things. J Netw Comput Appl 121:149–158
7. International Telecommunication Union (ITU) (2005) ITU internet reports 2005: the Internet of Things, executive summary. International Telecommunication Union (ITU), Geneva
8. Das A, Zeadally S, He D (2018) Taxonomy and analysis of security protocols for Internet of Things. Futur Gener Comput Syst 89:110–125
9. Raj A, Steingart D (2018) Review – power sources for the Internet of Things. J Electrochem Soc 165(8):B3130–B3136
10. Sharma A, Sharma P (2021) Energy harvesting technology for IoT edge applications. Smart Manufacturing-When Artificial Intelligence Meets the Internet of Things
11. Zeadally S, Das AK, Sklavos, N (2021) Cryptographic technologies and protocol standards for Internet of Things. Internet of Things 14:100075
12. Khan Y (n.d.) 5 Essential components of an IoT ecosystem. 21 Apr 2020. [Online]. Available: https://learn.g2.com/iot-ecosystem. Accessed 24 Jul 2022
13. Yugha R, Chithra S (2020) A survey on technologies and security protocols: reference for future generation IoT. J Netw Comput Appl 169:102763
14. Dilhac J-M, Boitier V (2016) 1 – Wireless sensor networks. In: Dilhac J, Boitier V (eds) Energy autonomy of batteryless and wireless embedded systems. Elsevier, London, pp 1–11

15. Suganya E, Sountharrajan S, Shandilya KS, Karthiga M (2019) Chapter 5 – IoT in agriculture investigation on plant diseases and nutrient level using image analysis techniques. In: Balas VE, Son LH, Jha S, Khari M, Kumar R (eds) Internet of Things in biomedical engineering. Academic Press, London, pp 117–130
16. Sanislav T, Mois G, Folea S, Miclea L (2016) Integrating wireless sensor networks and cyber-physical systems: challenges and opportunities. In: Control, robotics & sensors, 2016, cyberphysical system design with sensor networking technologies. IET Digital Library, London, United Kingdom, pp 47–76
17. Bello O, Zeadally S (2019) Toward efficient smartification of the Internet of Things (IoT) services. Futur Gener Comput Syst 92:663–673
18. Lazarescu MT (2017) Wireless sensor networks for the Internet of Things: barriers and synergies. In: Keramidas G, Voros N, Hübner M (eds) Components and services for IoT platforms. Springer, Cham. https://doi.org/10.1007/978-3-319-42304-3
19. Yue X et al (2017) Development of an indoor photovoltaic energy harvesting module for autonomous sensors in building air quality applications. IEEE Internet Things J 99:1
20. Lueth KL (2020) State of the IoT 2020. IoT analytics, 19 Nov 2020. [Online]. Available: https://iot-analytics.com/state-of-the-iot-2020-12-billion-iot-connections-surpassing-non-iot-for-the-first-time. Accessed 24 Jul 2022
21. E. P. & Technology (n.d) Energy harvesting for a greener environment

Microalgae Harvesting Strategies for Biofuel Production

Mayuri Gupta and Harsha Wakudkar

1 Introduction

Increasing global demand for energy results in fossil fuel usage. They are utilized by humans solely as the source of energy for food, electricity, and transportation. Still there is a large reservoir of fossil fuels available at reasonably low cost, although they are nonrenewable energy sources. It is available in limited supply and exhaustible in nature. The three major forms of nonrenewable energy sources are oil, coal, and natural gas. Its reserves in India as on 31.03.2014 stood at 594.49 million tons (MT), 319 billion tons (BT), and 1339.57 billion cubic meters (BCM), respectively. Fossil fuels dependence for meeting increasing energy demands is unsustainable due to lower production and increasing levels of consumption. In addition, burning of fossil fuels has caused number of environmental problems, such as emission of greenhouse gases resulting in global warming as well as imparting negative impacts on the environments and human health. It is estimated that energy-related GHG emissions will be double-fold by 2050 [1]. It is a matter of motivation for researchers to shift focus on the development of alternative energy sources which are solar, wind, and biomass. To limit the use of fossil fuels, these sources of renewable energy have been developed and widely used by different countries. Whatever energy source is used, it should not contribute to carbon emission; rather it should be as sustainable and carbon neutral as possible. A study conducted by the International Energy Agency (IEA) predicts that energy derived from renewables and waste has the highest potential. According to the report, hydro energy makes up 2.3% of TEPS, while others make up 1%, and bioenergy (including biomass and biofuels) makes up 10% of that (geothermal, solar, wind, heat) [1].

M. Gupta · H. Wakudkar (✉)
Central Institute of Agricultural Engineering, Bhopal, India

A. Nella et al. (eds.), *Energy Harvesting Trends for Low Power Compact Electronic Devices*, EAI/Springer Innovations in Communication and Computing,
https://doi.org/10.1007/978-3-031-35965-1_8

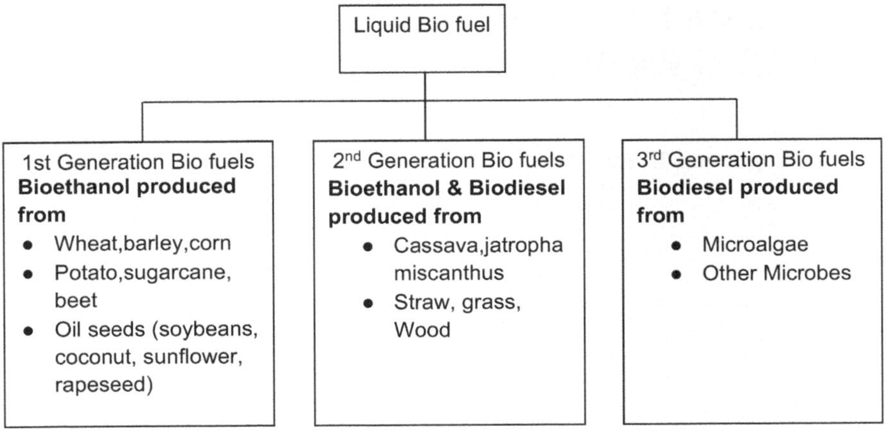

Fig. 1 Production sources of liquid biofuel

In contrast to other non-fossil fuel energy sources like wind and wave energy, biofuels are energy sources made from biological materials. The contentment of biofuels toward energy relies on the fact that the biological materials may contribute to the supply of energy, particularly liquid fuels for transportation emissions from which are the major contributors of greenhouse gas emissions to atmosphere. In contrast to fossil fuels, which take 10 to 100 million years to produce, biofuels, which can be solid, liquid, or gaseous, are self-sustaining and inexhaustible as they are made from livestock and plants which can be replaced in a very short amount of time. Solid biofuels, principally biomass, and gaseous biofuels mainly methane have been used for long span of time for cooking and energy and presently in combined heat and power systems and to generate electricity. It is the liquid biofuel that, when blended with petroleum diesel, can be used to substitute fossil fuels in current engines and transport infrastructure. They are produced from sugar, starch, cellulosic, or lipid-rich substrates. These agricultural residues include cereals like corn and wheat, sugar crops like sorghum and sugarcane, energy crops like switchgrass, agricultural wastes like straws and corn stover, domestic waste, and a number of aquatic species. Liquid biofuels can be divided into first- second-, and third-generation biofuels. First-generation biofuels are typically used to describe biofuels made from sugarcane and cereal crops (FGB). Energy sources made from lignocellulosic materials are known as second-generation biofuels (SGB). Third-generation biofuel production involves turning aquatic organisms like algae into liquid fuels (TGB) (Fig. 1).

There are significant economic, environmental, and political concerns with first-generation biofuels. The most concerning problem with first-generation biofuels is that more cultivable farmlands are required to produce first-generation biofuel raw material in order to increase production potential, which will lead to a reduction in the amount of land available for growing food for people and animals. The consequence of higher stress on agricultural land would lead to severe food shortages. The rising cost of food will undoubtedly push the poor and unprivileged out of their reach as the usage of first-generation biofuels increases.

Since first-generation biofuels are not very feasible, second-generation biofuels have been the focus of research. The main aim of this study is to create biofuels from lignocellulosic biomass, or the woody portion of plants that does not compete effectively with the human food chain [2]. According to Fig. 1, agricultural waste and residues, sawdust from the processing of wood, and nonedible parts of corn, sugarcane, beets, etc. are the main sources of second-generation biofuels (e.g., cuttings of plants, leaves, residues, wood cuttings, etc.). However, second-generation biofuels are fiscally profitless for large-scale production because conversion of plant/trees parts into fermentable substrates requires specialized and costly techniques to pretreat with certain enzymes [2, 3]. As a result, third-generation biofuels are the subject of most research. Microalgae is the primary ingredient in third-generation biofuels, as depicted in Fig. 1. Currently, it is regarded as a workable substitute renewable energy source for producing biofuels, overruling the drawbacks of first- and second-generation biofuels [2, 4–6].

2 Microalgae for Biofuel Production

2.1 Applicability of Microalgae for Biofuel

2.1.1 Microalgae

Photosynthetic microorganisms known as microalgae are the simplest type of plants found in nature in large quantities. Classification of algae is provided in five groups, namely, Chlorophyceae, Rhodophyceae, Phaeophyceae, Cyanophyceae, and Bacillariophyceae. Characteristics of every class are given in Fig. 2.

Despite having no rootstock, shoots, or leaves, microalgae have chlorophyll as their pigment for photosynthetic growth. They are able to transform atmospheric carbon dioxide, which serves as their main carbon source, into glucose for growth [7]. Microalgae, which represent a large diversity of species residing in a broad range of climatic conditions, are present in all current earth ecosystems including both terrestrial and aquatic. More than 50,000 species are thought to exist, but only a small fraction, roughly 30,000, have been investigated and analyzed [8].

2.1.2 Benefits of Using Microalgae for Biofuel Production

When compared to other materials, using microalgae for the generation of biodiesel has many benefits, according to numerous researchers [8]. Ideal characteristics for microalgae used as feedstock for biodiesel production include higher biomass productivities and effective lipid biosynthesis and productivity. Rapid growth encourages high biomass productivities, which in turn boosts yield per volume of harvested

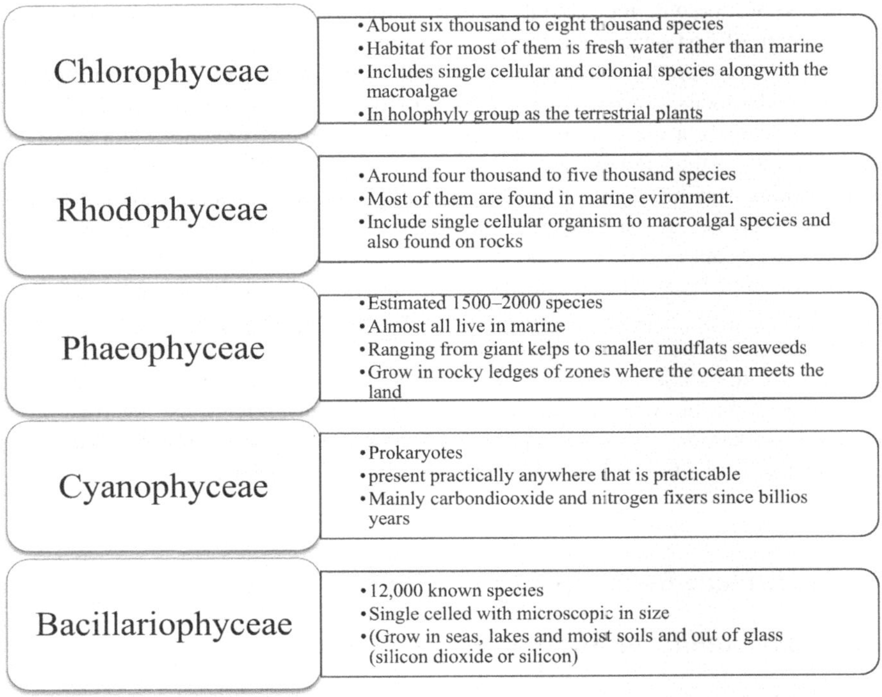

Fig. 2 Characteristics of different algae groups

material in a given time frame and lowers cost. In practice, they are easier to cultivate and can grow faster on minimum growth requirements than terrestrial plants. The main advantageous aspects of microalgae to be used for biofuel embrace the following: (1) They possess the capability to carry out oxygenic photosynthesis. (2) In contrast to particular land oilseed crops, they achieve maximum concentrations and have high levels of productivity per acre. (3) About 25–77% of dried biomass can make up the oil content and can produce 20,000 to 80,000 L of oil per acre per year which is 31 times more with respect to that generated by the prime terrene crop (palm tree). (4) They are not edible feed resources. (5) They make use of unusable and unarable land for cultivation. (6) They are able to grow in different types of water sources (fresh, brackish, seawater, and wastewater) [9]. (7) They produce both biofuels and valuable co-products.

However, biofuels cannot be fiscally competitive in the market because of its high expenses as compared to the fossil fuels; another viable way to reduce the cost of producing algal biofuel is to combine environmental pollution control measures, such as biological GHG (CO_2 and NOx) fixation and wastewater treatment to remove NH_4, NO_3^{2-}, and PO_4^{3-}. This will cause algae to utilize these nutrients and thus reduce pollution. Because of its high N:P ratio, the biomass leftover biomass after oil extraction can be converted into ethanol, methane, livestock feed, and organic fertilizer, or it can just be burned to produce energy for cogeneration (electricity and heat). Additionally, algae biodiesel reduces emissions of PM, CO, hydrocarbons, and SOx while performing as well as petro biodiesel and containing no sulfur.

2.1.3 Microalgae Lipid Content and Productivities

Microalgae are recognized as one of the earliest loving microbes on planet [10]. They multiply hundredth times quicker than the land plants, and their biomass is doubled in span of less than a day. In addition, a few varieties of microalgae can store a significant amount of lipid within the cellular structure, which can then be converted to biodiesel [11]. Microalgae consist of carbohydrates, proteins, and lipids. The lipid content varies among strains. Strains with little oil content consist of 30–50% protein, 25–40% carbohydrates, and 5–15% lipids (DW). Oleaginous algal cells contain 10–20% proteins, 25–30% carbohydrates, and 20–60% lipids (DW) [12]. Different strains of algae show different biochemical composition in terms of protein, carbohydrate, and lipid [13].

As biodiesel is generated by the trans-esterification of lipids (mainly triglycerides) with methanol (or other alcohol), higher lipid content is a desirable feature of using microalgae as biodiesel feedstock [14]. Table 1 shows that the yield of oil is higher with microalgae than that of vegetable oil crops which is also conferred by the strain [8]. It is also noted that though their oil contents are similar, microalgae are more advantageous for biodiesel production than other seed plants due to more overall biomass productivity in much low land use. In other words, cultivating microalgae for the generation of biodiesel necessitates the small land area and has significant advantages for efficient land use.

Generally speaking, the average oil content by dry weight of biomass for many microalgae ranges from 1 to 70%. Microalgae contain lipids and fatty acids in varying amounts depending on the cultivation conditions. By maximizing the

Table 1 Biochemical composition of algae expressed on a dry matter basis [13]

Strain	Protein (%)	Carbohydrates (%)	Lipids (%)
Anabaena cylindrica	43–56	25–30	4–7
Botryococcus braunii	40	2	33
Chlamydomonas reinhardtii	48	17	21
Chlorella vulgaris	41–58	12–17	10–22
Dunaliella salina	57	32	6
Euglena gracilis	39–61	14–18	14–20
Porphyridium cruentum	28–39	40–57	9–14
Prymnesium parvum	28–45	25–33	22–39
Scenedesmus dimorphus	8–18	21–52	16–40
Spirogyra sp.	6–20	33–64	11–21

Adopted from Lalman et al. [13]

Table 2 Comparison of microalgae with other biodiesel feedstocks [15]

Plant source	Seed oil content (% oil by wt in biomass)	Oil yield (L oil/ ha year)	Land use (m^2 year/kg biodiesel)	Biodiesel productivity (kg biodiesel/ ha year)
Corn/maize *(Zea mays L)*	44	172	66	152
Hemp *(Cannabis sativa L)*	33	363	31	321
Soybean *(Glycine max L)*	18	636	18	562
Jatropha *(Jatropha curcas L)*	28	741	15	656
Camelina *(Camelina sativa L)*	42	915	12	809
Canola/rapeseed *(Brassica annuus L)*	41	974	12	862
Sunflower *(Helianthus annuus L)*	40	1070	11	946
Castor *(Ricinus communis)*	48	1307	9	1156
Palm oil (Elaeis guineensis)	36	5366	2	4747
Microalgae *(low oil content)*	30	58,700	0.2	54,927
Microalgae *(medium oil content)*	50	97,800	0.1	86,515
Microalgae *(high oil content)*	70	136,900	0.1	121,104

Adopted from Rahman et al. [15]

growth-determining factors, it is possible to raise the lipid concentration by almost 90%. The amount of oil produced for each unit volume of microalgae culture per day is known as oil productivity, and it depends on both the rate of algal growth and the oil content of the biomass. Table 2 displays the oil content and different kinds of microalgae that can be found in freshwater and marine environments.

2.1.4 Differences Between Petro Diesel and Algae Biodiesel Attributes

- Algae biodiesel contains almost no sulfur.
- Because of its superior lubricating qualities, biodiesel extends the lifespan of injector machinery and reduces wear on fuel systems.

- Algae biodiesel will dissolve any remaining varnish residue because it has more potent solvent characteristics than petro diesel. To prevent clogging, fuel filters should be changed right away after adding biodiesel to systems that were previously powered by petroleum.
- Since biodiesel burns more efficiently and has improved lubricity to make up some of the energy density loss compared to petroleum diesel, the overall reduction in fuel efficiency is only about 2%.
- About 32 °F is the cloud point, or the temperature at which pure (B100) biodiesel begins to gel. In general, B20 blends (20% biodiesel, 80% petroleum diesel) do not gel in cold climates. The gel point of B100 will be lowered by various additives.
- The minimum temperature at which biodiesel can vaporize to form an extremely flammable mixture in air is 266 °F, which is much greater than the flash points of petroleum diesel (147 °F) and gasoline (52 °F).

Comparing biodiesel to petroleum diesel, particulate matter is reduced by almost 47%. Because biodiesel increases the oxygen content while decreasing the solid carbon fraction of the particulate matter, it produces less hazardous particulate matter (Table 3).

3 Microalgae Cultivation Systems

The growth conditions and species used have an impact on the structure and the rate of development of microalgae [17]. The processes and tools used to grow microalgae for biofuel are categorized using a variety of classification systems. Because of the special characteristics of microalgae, the significant scheme classifies the systems according to the nutritional source and the kinds of physiological processes that are used to increase algal biomass quickly. Based on this standard, cultures can be categorized mainly into four groups: photoautotrophic, heterotrophic, mixotrophic, and photoheterotrophic [18]. Table 4 shows the advantages and disadvantages of each type.

The two main types of microalgae farming systems are open and closed. The most well-liked of these systems are turf scrubbers and open ponds with lanes (raceway ponds). The most typical closed systems, also known as photobioreactors, are tubular, bubble column, airlift, and flat panel. Closed systems have a greater variety of shapes and configurations. Open ponds are the earliest and fundamental methods for producing microalgae. In general, these methods consist of an impermeable oval pond with a continuous or divided form and a series of plates through which water is circulated or transferred to facilitate uniform mixing of nutrients and algal cells [20]. Raceway ponds and similar systems differ from open ponds in that they have artificial agitation mechanisms, which are typically performed by paddle impellers [21].

Table 3 Typical ranges of lipid content and productivity of selected marine and freshwater [16]

Microalga species	Lipid content (%, w/w_{DW})	Lipid productivity (mg L^{-1} d^{-1})	Natural habitat
Botryococcus spp.	25.0–75.0	–	Freshwater
Chaetoceros calcitrans	14.6–39.8	17.6	Freshwater
Chaetoceros muelleri	33.6	21.8	
Chlorella emersonii	25.0–63.0	10.3–50.0	Freshwater
Chlorella protothecoides	14.6–57.8	1214	
Chlorella pyrenoidosa	2.0	–	
Chlorella sorokiniana	19.0–22.0	44.7	
Chlorella vulgaris	5.0–58.0	11.2–40.0	
Chlorella spp.	10.0–57.0	18.7–42.1	
Chlorococcum spp.	19.3	53.7	Freshwater
Dunaliella primolecta	23.1	–	Freshwater
Dunaliella salina	6.0–25.0	116.0	
Dunaliella tertiolecta	16.7–71.0	–	
Dunaliella spp.	17.5–67.0	33.5	
Ellipsoidion spp.	27.4	47.3	Freshwater
Haematococcus pluvialis	25.0	–	Freshwater
Isochrysis galbana	7.0–40.0	–	Seawater
Isochrysis spp.	7.1–33.0	37.8	
Nannochloris spp.	20.0–56.0	60.9–76.5	Seawater
Nannochloropsis oculata	22.7–29.7	84.0–142.0	Seawater
Nannochloropsis spp.	12.0–53.0	60.9–76.5	
Neochloris oleoabundans	29.0–65.0	90.0–134.0	Seawater
Pavlova salina	30.9	49.4	Seawater
Pavlova lutheri	35.5	40.2	
Phaeodactylum tricornutum	18.0–57.0	44.8	Seawater
Scenedesmus obliquus	11.0–55.0	–	Freshwater
Scenedesmus quadricauda	1.9–18.4	35.1	
Scenedesmus sp.	19.6–21.1	40.8–53.9	
Spirulina platensis	4.0–16.6	–	Seawater

Adopted from Mishra and Ojha [16]

Closed cultivation methods, also known as closed photobioreactors (PBRs), are more effective when quality is considered because they can be functioned under extremely controlled circumstances, overcoming the disadvantages of open cultivation systems. Photobioreactors are developed to grow isolates of photoautotrophic that are abundant in high-value products; by correctly adjusting the culture conditions, the production volume of the final products can be increased. Closed

Table 4 Comparison of features and challenges of algal cultivation methods [19]

Type of culture	Photoautotrophic	Heterotrophic	Mixotrophic	Photoheterotrophic
Type	Closed and open systems	Close photobioreactor	Photobioreactor	Photobioreactor
Advantages	*Closed*: Water saving, greater long-term culture maintenance, high yield *Open*: Lower costs, evaporative cooling	Easy to maintain, high biomass concentrations, contamination prevention, utilization of inexpensive lignocellulosic sugars	Two-route cultivation, easy to bioremediate, uses organic compounds as energy source, provides superior energy recovery and carbon footprint	Fast growth of algae and synthesis of valuable metabolites (i.e., fatty acids)
Issues	*Closed*: High-cost, temperature control (requires cooling), maximum light exposure, periodic cleaning *Open*: Low yield, changes in humidity and temperature, maximum light exposure	Competition with other biofuel technologies for feedstock	Design needs to be upgraded to improve operation and economy, rarely used for bio-oil production, not used for biodiesel production	Requires light as an energy source, unlike mixotrophic cultivation, not used for biodiesel production

Adopted from Debowski et al. [19]

photobioreactor designs must be tailored specifically for the alga of interest, according to their physiological characteristics and specific growth [22]. Photobioreactors, as opposed to open cultivation system, reduce contamination and permit microalgal monoculture while managing factors like pH, temperature, light, and CO_2 concentration. These systems result in lower CO_2 and water evaporation losses, higher CO_2 mass transfer rates, and higher complexity bioproduct production [23].

Depending on the requirements, the precepts and design of cultivation systems change [24]. In wastewater treatment facilities, open ponds may be circular or gravity-fed. Similar to this, the fundamental tubular design of PBRs has been upgraded in the last 10 years to enable fair light penetration and blending of culture, enabling the manufacturing of different medicinal and value-added dietary products.

Although open and closed cultivation techniques have been the subject of extensive research, two-stage composite cultivation method has received less attention. Due to their capability to efficaciously distinct the biomass growth period from the lipid accumulation phase, two-phase composite cultivation method has been suggested as preferable microalgae culture method [24, 25].

4 Harvesting Strategies

Despite the many uses, microalgae biomass production is currently not profitable. To increase the process' economics, a number of research directions have been looked into. The wastewater can be used as a media for growth of microalgae which leads to lessening of operational costs and the amount of freshwater required. By claiming credit for all commercially valuable products made from microalgae, a manufacturing approach based on biorefineries lowers total costs of production [26]. Due to the fact that this production stage adds on the 20–30% expense in the total production cost, it is important to look at harvesting methods lower in costing [27]. The major causes of higher operational costs are the smaller dimensions of microalgae and the growth of those organisms in extremely diluted cultures (mass concentration less than 1 g/L) with densities that are comparable to those of water [28, 29]. The cells also include algogenic organic matter, which preserves the microalgal cells' scattered state, and the surface of the algae is negative in charge [30]. There is currently no economically viable and efficient microalgal harvesting method. Thus, it is believed that reducing harvesting costs will be essential for the long-term mass production of microalgal biomass [27]. Higher concentration of biomass and minimal operational, energy, and maintenance costs are desirable characteristics of an ideal harvesting procedure for the majority of microalgae species [28]. It is customarily expensive to start the harvesting process because it impacts how things will be processed afterward. Generally either of the two methods were used: a two-step concentrating process in which the suspension is concentrated to a slurry having 2–7% total suspended solids, TSS, and then dehydrated to a cake containing 15–25% TSS or a one-step concentration methodology [31, 32]. Moreover, when selecting a suitable harvesting process, keep in mind that microalgae mass needs to be treated further. As a result, these processes must be not harmful and not cause contamination microalgae biomass. Additionally, it is advantageous if the culture medium may be recycled using the selected harvesting method [32]. Microalgal harvesting usually consists of a two-step concentrating process: thickening and dewatering. These steps are essential for separating the preliminary suspension of algae into a dense algal slurry and enabling additional downstream processing. Microalgal harvesting now uses physical, chemical, biological, and, to a lesser degree, electric approaches. Combining two or more of these methods is very common in order to achieve a higher recovery rate at a low cost. In reality, combining centrifugation with flocculation-sedimentation can dramatically reduce operational expenses [33].

4.1 *Centrifugation*

The fastest harvesting method is centrifugation; however because of its large energy demand and increased cost, it is only used for extremely valuable items like pharmaceutical drugs, highly unsaturated fats, and other products

[34]. Centrifugation leads to the higher algal recovery up to 94% and with the energy input 20 kWh m^{-3}, which is very high when compared with 17% algal recovery with energy input of 0.80 kWh m^{-3} [35]. Centrifugation is a concentrated biomass dewatering method with a higher life cycle energy demand (700 MJ/dry tone of algae), greenhouse gas emissions (50 kg CO_2 eq./dry tone of algae) [36], and capital cost than other harvesting method. The vast majority of microalgae can be harvested using centrifuges. Usually, centrifuges are configured to enhance extraction effectiveness. However, even the most economical method of harvesting microalgae might not be the most effective [37]. Centrifugal force can be used to recover most microalgae from a diluted dispersion. However, it may still be considered a decent option, particularly in the microalgae cultivation for value-added products or when used in conjunction with other initial harvesting techniques [36]. To use this strategy for microalgae cultivation, energy-efficient methods for concentrating biomass before centrifugation must be developed [38].

4.2 Coagulation and Flocculation

Coagulation or flocculation is typically accompanied by gravitational settling for low-cost microalgae harvesting [39]. In contrast to coagulation, which calls either adjusting the pH or adding salts, flocculation is accomplished by adding charged polymeric materials to the algal culture. Consequently, the definition of chemical coagulation/flocculation is the combination of finely dispersed particles that are suspended onto bigger aggregation followed by the conglomeration of these into greater globules that sink to the bottom of the container, generating a clean supernatant [40]. Various methods that can be used to get follicles to form are (i) bridging, which happens when polymeric materials or colloidal particles bind to the edges of two distinct particles a connection between them; (ii) electrostatic patch (or patching), which happens when a positive-charged polymer attaches to a negatively charged particle or vice versa, indigenously changing that charge and creating a patch that will connect with oppositely charged patches; and (iii) sweep flocculation, which happens when particles are trapped in a large mineral precipitation [41]. Flocculation is the formation of massive assemblages from poorly connected suspensions of algal cells. The relationships between a flocculation chemical and the surface charge of the suspended cells cause them to initially aggregate into bigger particles. Following that, the aggregates coalesce into large flakes and separate from the suspension. The flocculation technique is thought to be an efficient way of quickly processing large amounts of biomass. Various chemicals, including inorganic salts of multivalent metal species and organic polymers, have been tested as flocculating agents (polyelectrolytes) [42]. Microorganisms have awhile back been used in the flocculation of some microalgae species [43]. Operations involving physical processes, such as gravitational settling, flotation, centrifugation, or filtration/screening, usually require a preliminary stage of flocculation for cellular grouping. The use of inorganic coagulants produces a large volume of sludge, causing microalga death or growth inhibition. Although some microalgae species respond well to aluminum as a coagulant, using metal salts may not be appropriate if the biomass is going to be utilized for fish farming, animal feed, or fertilizer. Ideally, flocculating agent should be cheap, safe, and effective at minor concentration [44]. Ferric chloride, aluminum sulfate ($Al_2(SO_4)_3$), and ferric sulfate ($Fe_2(SO_4)_3$) are some of the popular coagulating agents. The flocculation efficiency with different coagulating agents, viz., ferric chloride, sodium hydroxide, and alum, was observed in the range of 15–94%. Although metal coagulation agents like alum and ferric chloride can quickly flocculate the material, it is quite costly because vast quantities of these chemicals are needed. Additionally, the employed metals may pollute algal biomass that were treated with these chemical species, preventing its use as a biofuel or an animal fodder.

4.3 Electrochemical Methods

Electrochemical methods are physiochemical procedures that dissolve sacrificial metal electrodes to produce the ions required for coagulation/flocculation. Electrochemical methods, as opposed to chemical flocculation, do not introduce anions that would result in pH reduction [45]. However, these techniques are adaptable because they may be used with a variety of microalgae cells and are environmentally beneficial (no chemicals are required) [32, 46]. Because cells of microalgae possess negative charge, they can be segregated by giving an electric current to the microalgae culture [32]. They can be deposited on the electrodes (electrophoresis) and deposit at the bottom of the container (electro-flocculation). The only aspect of the electro-flotation technique that differentiates from DAF is how bubbles are generated. Water electrolysis produces hydrogen bubbles, which seems to be less expensive than conventional DAF [47]. This procedure is known as electro-coagulation-flotation (ECF). There are two kinds of electrode that can be used: sacrificing and non-sacrificing. Metal ions are released into the broth as a result of the use of sacrificial electrodes, which in turn depends upon the quantity of electric current that is given to the electrolytic solution. The dissolution of the reactive anode's emits ions results in the formation of coagulant, which destabilizes the microalgal suspension [48]. As a result, microalgae cells get aggregated by interacting with positive charge metallic ions. The toxicity of this electrode dissolution cannot be compared with the introduction of chemical coagulants into the microalgae culture. As a result, in both the methods, the likelihood of metallic hydroxide contamination risk is minimal. The usage of a non-sacrificial electrodes is dependent solely on the progression of negative charge microalgal cells toward anode. When cells reach to anode, their charge gets loosed and they aggregate. As a result, when compared to chemical methods, this kind of device is more likely to foul. An iron electrode just attained a removal of 78.9% during a 45-min harvesting period, compared to an aluminum electrode's 100% total extraction efficiency, primarily due to the low current efficiency and lower degree of dissociation of iron electrodes. As a result, ferric electrodes are less used than aluminum electrodes because they use much energy and produce slurry that is brown in color as a result of the formation of ferric oxide [48].

An attempt was made to assess various harvesting methodologies, and on the basis of performance and energetics, the ECF method was further tested and analyzed for harvesting of microalgae. The maximum harvesting efficiency of electro-flocculation system was achieved up to 93.1% with average energy consumption of 2.715 kWh/kg. The integration of solar PV panels of 270 W and 31.50 V may further reduce the harvesting cost and make it more feasible process for harvesting of microalgae.

4.4 Gravity Sedimentation

Despite its simplicity, sedimentation is extremely energy-efficient and effective for a variety of microalgae [26]. Thus, gravitational settling should be used for microalgal harvesting when the final use of harvested microalgae does not require high purity, such as biofuels [9, 29]. The confidentiality of this method is low because microalgal density is essential to the effectiveness of the processing. Microalgae cells sedimentation rate of 0.1–2.6 cm h^{-1} leads to a very slow sedimentation process, which in turn results in the deterioration of microalgae biomass and limited the usage of this methodology for regular harvesting [27]. To accelerate microalgal settling, a coagulation or flocculation step is applied before the gravitational settling. Although using deposition tanks is thought to be a straightforward and affordable method, the purity that may be obtained without prior coagulation or flocculation is quite low. Similar to this, lamella-type separators' ability to concentrate microalgae is low and uncertain, necessitating additional thickening [48].

4.5 Filtration

Filtration is the most cost-effective method of harvesting, and it is possible to concentrate microalgae with efficiencies ranging from 10 to 20% (w w^{-1}), though the process necessitates regular maintenance, such as filter cleaning and renewal [49]. Available filtration techniques include microfiltration, ultrafiltration, pressure filtration, vacuum filtration, and tangential flow filtration. Pressure or vacuum filters can be used to concentrate microalgae strains with large cell dimensions, like *S. platensis*. The recovery of microalgae such as *Chlorella* and *Dunaliella*, with cell size similar to those of bacteria, cannot be carried out by methods involving pressure or vacuum filtrations [50, 51].

5 Conclusion

Microalgae systems-based technologies have proven to be superior to physical, chemical, and biological methods used in traditional treatment of wastewater and gaseous effluent. Microalgae have basic cell structures, are able to thrive and procreate in almost any surroundings, and can withstand a broad spectrum of environmental factors and pollutants. They only need water, light, and CO_2 to grow. There is no one technique that can be employed to effectively harvest all microalgal strains. Based on microalgal characteristics like cell morphology and size, cell surface characteristics, culture medium characteristics, and the quality and worth of the final product, an effective method should be developed. Furthermore, a combination of processes is commonly used to improvise harvesting efficiency and costing. A two-stage separation typically involves thickening and dewatering procedures. In terms of biofuel application, bio-flocculation accompanied by gravitational sedimentation appears to be a practical way to recover microalgae cells at a reasonable price. However, the potential for microbial contaminations may restrict the use of bio-flocculation when the separation is intended to produce animal feed or high-value compounds. Electrical-based techniques may be an option for avoiding the toxic effects of chemical coagulants or flocculants as well as microbiological contaminations from bio-flocculation processes.

References

1. Energy efficiency market report (2013) International Energy Agency
2. Dragone G, Fernandes B, Vicente AA, Teixeira JA (2010) Third generation biofuels from microalgae. In: Mendez-Vilas A (ed) Current research, technology and education topics in applied microbiology and microbial biotechnology. Formatex, Badajoz, pp 1355–1366
3. Brennan L, Owende P (2010) Biofuels from microalgae – a review of technologies for production, processing, and extractions of biofuels and coproducts. Renew Sust Energ Rev 14:557–577
4. Nigam PS, Singh A (2011) Production of liquid biofuels from renewable resources. Prog Energy Combust Sci 37(1):52–68
5. Chisti Y (2007) Biodiesel from microalgae. Biotechnol Adv 25:294–306
6. Li Y, Horsman M, Wu N, Lan CQ, Dubois-Calero N (2008) Biofuels from microalgae. Biotechnol Prog 24:815–820
7. Ho SH, Chen CNN, Lai YY, Lu WB, Chang JS (2014a) Exploring the high lipid production potential of a thermotolerant microalga using statistical optimization and semi-continuous cultivation. Bioresour Technol 163:128–135. https://doi.org/10.1016/j.biortech.2014.04.028
8. Mata TM, Martins AA, Caetano NS (2010) Microalgae for biodiesel production and other applications: a review. Renew Sust Energy Rev 14:217–232. https://doi.org/10.1016/j.rser.2009.07.020
9. Tamagnini P, Leitao E, Oliveira P, Ferreira D, Pinto F, Harris DJ (2007) Cyanobacteria hydrogenase: diversity, regulation and applications. FEMS Microbiol Rev 31:692–720
10. Song D, Fu J, Shi D (2008) Exploitation of oil-bearing microalgae for biodiesel. Chin J Biotechnol 34:341–348
11. Chisti Y (2008) Biodiesel from microalgae beats bioethanol. Trends Biotechnol 26:126–131

12. Hu Q, Sommerfeld M, Jarvis E, Ghirardi M, Posewitz M, Seibert M, Darzins A (2008) Microalgal triacylglycerols as feed-stocks for biofuel production; perspectives and advances. Plant J 54:621–639
13. Lalman JA, Shewa WA, Gallagher J, Ravella S (2016) Biofuels production from renewable feedstocks. In: Lau PCK (ed) Quality living through chemurgy and green chemistry. Springer, New York, pp 193–220
14. Liu J, Qiang H (2013) Chlorella: industrial production of cell mass and chemicals. In: Richmond A, Qiang H (eds) Handbook of microalgal culture. Wiley, Oxford, pp 327–338
15. Rahman A, Prihantini NB, Nasruddin (2019) Fatty acid of microalgae as a potential feedstock for biodiesel production in Indonesia. In: AIP conference proceedings 2062, 020059
16. Mishra S, Ojha S (2013) Potential of oilgae: the prospects and challenges. In: Microbiology applications. Har Krishan Bhalla & Sons, Dehradun
17. Aziz MMA, Kassim AK, Shokravi Z, Jakarni FM, Liu HY, Zaini N (2020) Two-stage cultivation strategy for simultaneous increases in growth rate and lipid content of microalgae: a review. Renew Sust Energ Rev 119:109621
18. Piasecka A, Nawrocka A, Wiacek D, Krzeminska J (2020) Agro-industrial by-product in photoheterotrophic and mixotrophic culture of Tetradesmus obliquus: production of ω3 and ω6 essential fatty acids with biotechnological importance. Sci Rep 10:1–11
19. Marcin D, Marcin Z, Joanna K, Natalia K, Szymon T (2020) Microalgae cultivation technologies as an opportunity for bioenergetic system development – advantages and limitations. Sustainability 12:9980. https://doi.org/10.3390/su12239980
20. Handler RM, Canter CE, Kalnes TN, Lupton FS, Kholiqov O, Shonnard DR, Blowers P (2012) Evaluation of environmental impacts from microalgae cultivation in open-air raceway ponds: Analysis of the prior literature and investigation of wide variance in predicted impacts. Algal Res 1:83–92
21. Ketheesan B, Nirmalakhandan N (2011) Development of a new air-lift-driven raceway reactor for algal cultivation. Appl Energy 88:3370–3376
22. Sierra E, Acien FG, Fernandez JM, García JL, Gonzalez C, Molina E (2008) Characterization of a flat plate photobioreactor for the production of microalgae. Chem Eng J 138:136–147
23. Singh RN, Sharma S (2012) Development of suitable photobioreactor for algae production – a review. Renew Sust Energ Rev 16:2347–2353
24. Schenk PM, Thomas-Hall S, Stephens E, Marx U, Mussgnug J, Posten C et al (2008) Second generation biofuels: high-efficiency microalgae for biodiesel production. Bioenergy Res 1:20–43. https://doi.org/10.1007/s12155-008-9008-8
25. Su C-H, Chien L-J, Gomes J, Lin Y-S, Yu Y-K, Liou J-S et al (2011) Factors affecting lipid accumulation by *Nannochloropsis oculata* in a two-stage cultivation process. J Appl Phycol 23: 903–908. https://doi.org/10.1007/s10811-010-9609-4
26. Rawat I, Ranjith Kumar R, Mutanda T, Bux F (2011) Dual role of microalgae: phycoremediation of domestic wastewater and biomass production for sustainable biofuels production. Appl Energy 88:3411–3424
27. Christenson L, Sims R (2011) Production and harvesting of microalgae for wastewater treatment, biofuels, and bioproducts. Biotechnol Adv 29:686–702
28. Danquah MK, Gladman B, Moheimani N, Forde GM (2009) Microalgal growth characteristics and subsequent influence on dewatering efficiency. Chem Eng J 151:73–78
29. Molina Grima E, Belarbi EH, Acién Fernández FG, Robles Medina A, Chisti Y (2003) Recovery of microalgal biomass and metabolites: process options and economics. Biotechnol Adv 20:491–515
30. Danquah MK, Ang L, Uduman N, Moheimani N, Forde GM (2009) Dewatering of microalgal culture for biodiesel production: exploring polymer flocculation and tangential flow filtration. J Chem Technol Biotechnol 84:1078–1083
31. Brennan L, Owende P (2010) Biofuels from microalgae – a review of technologies for production, processing, and extractions of biofuels and co-products. Renew Sust Energ Rev 14:557–577

32. Uduman N, QiY DMK, Forde GM, Hoadley A (2010) Dewatering of microalgal cultures: a major bottleneck to algae based fuels. J Renew Sustain Energy 2:012701–012715
33. Schlesinger A, Eisenstadt D, Bar-Gil A, Carmely H, Einbinder S, Gressel J (2012) In expensive non-toxic flocculation of microalgae contradicts theories; over- coming a major hurdle to bulk algal production. Biotechnol Adv 30:1023–1030
34. Zhou W, Min M, Hu B, Ma X, Liu Y, Wang Q (2013) Filamentous fungi assisted bio-flocculation: a novel alternative technique for harvesting heterotrophic and autotrophic microalgal cells. Sep Purif Technol 107:158–165
35. Barros AI, Gonçalves AL, Manuel S, Pires José CM (2015) Harvesting techniques applied to microalgae: a review. Renew Sust Energ Rev 41:1489–1500
36. Udom I, Zaribaf BH, Halfhide T, Gillie B, Dalrymple O, Zhang Q, Ergas SJ (2013) Harvesting microalgae grown on wastewater. Bioresour Technol 139:101–106
37. Dassey AJ, Theegala CS (2013) Harvesting economics and strategies using centrifugation for cost effective separation of microalgae cells for biodiesel applications. Bioresour Technol 128: 241–245
38. Salim S, Vermue MH, Wijffels RH (2012) Ratio between auto-flocculating and target microalgae affects the energy-efficient harvesting by bio-flocculation. Bioresour Technol 118: 49–55
39. Smith BT, Davis RH (2012) Sedimentation of algae flocculated using naturally-available, magnesium-based flocculants. Algal Res 1:32–39
40. Papazi A, Makridis P, Divanach P (2010) Harvesting Chlorella minutissima using cell coagulants. J Appl Phycol 22:349–355
41. Vandamme D, Foubert I, Muylaert K (2013) Flocculation as a low-cost method for harvesting microalgae for bulk biomass production. Trends Biotechnol 31:233–239
42. Kim D-G, La H-J, Ahn C-Y, Park Y-H, Oh H-M (2011) Harvest of Scenedesmus sp. with bioflocculant and reuse of culture medium for subsequent high-density cultures. Bioresour Technol 102:3163–3168
43. Wu Z, Zhu Y, Huang W, Zhang C, Li T, Zhang Y, Li A (2012) Evaluation of flocculation induced by pH increase for harvesting microalgae and reuse of flocculated medium. Bioresour Technol 110:496–502
44. Ahmad AL, Mat Yasin NH, Derek CJC, Lim JK (2011) Optimization of microalgae coagulation process using chitosan. Renew Sust Energ Ver 173:879–882
45. Lee AK, Lewis DM, Ashman PJ (2013) Harvesting of marine microalgae by electro-flocculation: the energetics, plant design, and economics. Appl. Energ 108:45–53
46. Zenouzi A, Ghobadian B, Hejazi M, Rahnemoon P (2013) Harvesting of microalgae Dunaliella salina using electroflocculation. J Agric Sci Technol 15:879–888
47. Show K-Y, Lee D-J (2014) Algal biomass harvesting. In: Pandey A, Lee D-J, Chisti Y, Soccol CR (eds) Biofuels from algae. Elsevier, Burlington, pp 85–110
48. Barros AI, Gonçalves AL, Simões M, Pires JCM (2015) Harvesting techniques applied to microalgae: a review. Renew Sust Energ Rev 41:1489–1500
49. Weschler MK, Barr WJ, Harper WF, Landis AE (2014) Process energy comparison for the production and harvesting of algal bio-mass as a biofuel feedstock. Bioresour Technol 153:108–115
50. Harun R, Singh M, Forde GM, Danquah MK (2010) Bioprocess engineering of microalgae to produce a variety of consumer products. Renew Sust Energ Rev 14:1037–1047
51. Pacheco MM, Hoeltz M, Moraes MS, Schneider RC (2015) Microalgae: cultivation techniques and wastewater phycoremediation. J Environ Sci Health 50(6):585–601

Energy Harvesting Scheme Using Queuing Theory for Wireless Body Area Network

R. Nidhya, D. Pavithra, R. Kalpana, M. Kathirvelu, and P. Jayarajan

1 Introduction

Monitoring exact indicators in several applications like atmosphere observing, habitant examining, combat sector, farming pasture observing, and perceptive homes are using in wireless sensor networks (WSNs). Detecting region to screen field is dispersed by these remote sensors. Develop a subfield of WSN by WBAN. Well-being checking of WBAN is the key operation. Sensors are placed on the body or entrenched in the body to monitor the vital signs like heart rate, circulatory strain, glucose level, body temperature, and so on [1]. To make use of WBAN to monitor well-being factor drastically decreases the operations of patient in curing facility [3, 10]. For more periods, with the aid of WBAN innovation assistance, patients are monitored at home. Sensors interminably sense data which are promoted to medical server. Sensor hubs are sunk with condition-based vitality cause in WBANs. It is necessary to make use of tiny amount of power for transferring data from sensor hubs to sink. To renew the batteries is one of the real barriers in WBAN [8, 9]. To

R. Nidhya (✉) · R. Kalpana
Department of Computer Science and Engineering, Madanapalle Institute of Technology & Science, Madanapalle, Andhra Pradesh, India

D. Pavithra
Department of Information Technology, Dr. NGP Institute of Technology, Coimbatore, Tamil Nadu, India

M. Kathirvelu
Department of ECE, KPR Institute of Engineering and Technology, Coimbatore, Tamil Nadu, India
e-mail: kathirvelu.m@kpriet.ac.in

P. Jayarajan
Department of ECE, Sri Krishna College of Technology, Coimbatore, Tamil Nadu, India

A. Nella et al. (eds.), *Energy Harvesting Trends for Low Power Compact Electronic Devices*, EAI/Springer Innovations in Communication and Computing,
https://doi.org/10.1007/978-3-031-35965-1_9

beat this concern of energizing batteries is necessary for a productive steering. In WSN innovation, numerous vitality effective directing conventions are planned. However, WSNs and WBANs have distinct models, applications, and work in various circumstances [5].

To port WSN steering conventions to WBAN is difficult. To monitor patients for more periods, vitality effective directing convention for WBAN is mandatory [4]. We propose a high reliable, throughput, and stable steering convention for WBAN. Sensors for ECG as well as glucose level are placed close to the sink. In this way both sensors contain the primary information of patient, and it is necessary for least weakening, high firm quality, and long life; and their information specifically to sink are dependably transmitted through these sensors. After their security hub, various sensors take and broadcast their data to sink through transmitter hub. For more periods, it contains the auxiliary vitality of hubs and system works. Transmit, receive, idle, and sleep are the four nodes which operate a typical wireless radio. For the duration of sleeping, maximum power is consumed in transmission. Circuits can be turned off to some extent during energy efficiency [6, 7]. The balanced energy consumption [BEC] protocol works based on the cost function calculated for each relay node in network. To balance the load, based on the cost function, relay nodes are rotated after each round [2]. The queuing system is defined as follows: a set of customers are waiting for certain service, but providing services is not possible due to various reasons. Real-life examples are printing more pages simultaneously in a printer, car assembly system, message transmission, etc. Queuing theory is utilized to optimize the system performance by reducing the cost of operation [11]. But there is a minimum requirement system should meet for queuing theory. This proposed work used to evaluate performance analysis for remote health monitoring system using queuing models.

The chapter is organized as follows: Sect. 2 describes the related works of the proposed system, Sect. 3 explains the WBAN architecture, Sect. 4 discusses about proposed methodology, Sect. 5 presents modeling of WBAN using queuing theory, Sect. 6 analyzes the results and discussion of proposed methodology, and Sect. 7 finalizes the conclusion and future enhancement.

2 Related Work

A WBAN is a set of sensors implanted or wearer in body to monitor the vital parameters of human remotely using Bluetooth or Wi-Fi or Internet [12–14]. The wearable type of devices placed on skin or close to the body of the patient [15] similarly implantable devices can be fixed inside of the body to monitor the patient health parameters such as blood pressure, heart rate, electrocardiography (ECG), etc. remotely [16, 17]. Reliable and on-time information delivery in WBAN is the most important but that leads to many challenges. There are many researchers working on reliable data delivery in WBAN [18]. In paper [19] authors proposed the weighted fair queuing model for better packet scheduling in wireless communication systems, and proposed model is the energy-efficient model also. Few researchers have proposed analytic solution model for rectifying queuing delays in MAC layer of wireless nodes [20]. To enhance and analyze the performance of ad hoc wireless network, researchers have used G/G/1 queuing system. But this proposed solution did not provided the better QoS because the system produced only approximate results and unrealistic traffic assumptions.

Nageswari et al. in [21] have performed the analysis of cluster-based homogeneous sensor network using energy-efficient N-policy (EENP) model to explore the improvement in energy efficiency using a threshold model. The authors in [22, 23] have proposed NEEF and EABM algorithm for WBSN which describes the energy minimization scheme during transmission of data. The authors in [24–27] have done extensive research in priority-based queueing models for WSN applications which helps to transmit the high-priority data with very less delay but still incorporating the queue threshold model. The authors have also considered the node failure and channel contention in analyzing the performance of the network to highlight the impact with respect to energy. Ashraf et al. in [28] have explored the node lifetime estimation of the network based on queue threshold model.

In paper [29] the authors proposed the enhanced energy-aware quality of service routing protocol for wireless sensor network. This method supports two different types of traffics. These two different traffics can access each other's bandwidth. By considering the cost function of each link, the proposed routing protocol provides the output data with low cost and less delay in path. In [30] authors proposed an analytical framework which utilizes the M/G/1 queuing system to assure the QoS in WSN. Authors also used Poisson distribution model to analyze traffic behavior in WSN. Unfortunately Poisson model is not able to handle the traffic in data communication. So authors proposed new framework, G/M/1 queuing system for multimedia sensors. But this method extracts only analytical result. So enhanced version is proposed in paper [31]. In this method authors extended G/M/1 queuing system with priority-based model to analyze the three queues to handle multiple traffics classes at the same time. Usually the WBAN traffic is either application driven or event driven. The proposed methodology ensures the QoS enhancement for handling multiple classes of traffic WBAN input.

3 Architecture of WBAN

The human body is attached by the WAN. The WAN contains a wireless sensor and is responsible for communicating, processing, and sensing the physiological signal to the medical server for more analysis. From hospital information system (HIS), sensors share information to the medical server with the help of diverse networking strategies depending on their distance. Monitoring of physiological changes in human body using wireless sensors can be invasive or noninvasive. Personal area network (PAN), local area network (LAN), and wide area network (WAN) are the different network strategies used in WBAN.

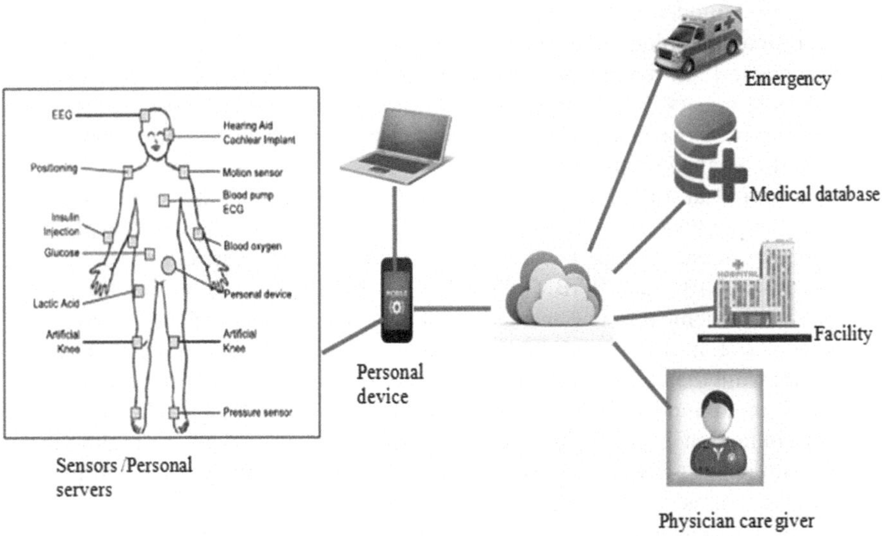

Fig. 1 Patient-centered body area network

Three-tier network for communication is the standard architecture of WBAN. The sensors attached to body either invasively or noninvasively are the first-tier architecture. On/in the body has one or more sensors. At a cluster head, two or more sensors connected to one patient make a cluster and the data from cluster is grasped. PAN coordinator sends data by the cluster head. At first tier communication network ZigBee, MICS, and Bluetooth are used [2].

In second tier, LAN is existing. Sending the information is from PAN coordinator to the LAN router. Wi-Fi access point is an example. In third tier instead of using LAN and WAN technologies, Ethernet or personal digital assistant (PDA) with Internet facility is used. WBANs are classified as patient-centered WBAN (Fig. 1) and hospital-centered WBAN as stated by their use. Patient-centered WBAN systems are usually used for distant health monitoring. They are classified by a three-tier architecture consisting of sensors on the body or implanted in the body of the patient for monitoring of necessary parameters; a hospital information system (HIS) for storage and management of health information, which is generally situated in the medical facility; and a mobile device that generally acts as a personal entry connecting the two previous entities. Figure 1 depicts the architecture of patient-centered body area network.

4 Proposed Methodology

An addition of wireless sensor network is wireless body area network. In this method, WBAN sensor nodes have been located on the human body parts to sense different types of data. Sink node transmits the sense information over the network. Based on routing protocol, sensor node transmits the information for data broadcast over the network. In the proposed work, various nodes have been used for information forwarding and information aggregation across the network, and different sensors have been posted over the network which are mainly used for sensing the data. For better energy consumption throughout the network, these nodes required route selection. For network lifetime, energy is the main limitation. On chosen path of the system, energy awareness has to be accomplished. Energy-aware cluster routing method has been used in the proposed work for network deployment [12, 13] (Fig. 2).

A. ***Parameter Setup***

Various indicators have been initialized for network deployment and configuration in this phase of networking. These parameters are vital for nodes of pattern and nodes data spread over the network. The below table corresponds to a mixture of simulation parameters that have been defined in the proposed work. In the proposed work, these different parameters are used for network initialization and data transmission [11] (Table 1).

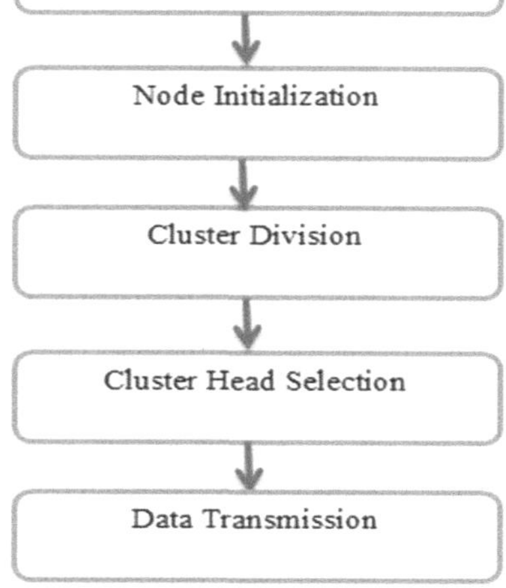

Fig. 2 Flow of proposed work

Table 1 Simulation parameter setup

Parameters	Value
Total number of nodes (n)	12
Transmission energy (ETX)	50 nJ
Receiving energy (ERX)	50 nJ
Aggregation data energy (EDA)	50 nJ
Total energy (E)	0.5
Frequency (F)	2.54 GHz
Message size (S)	4000 bites

B. ***Deployment of Nodes***

After the initialization of diverse indicators of simulation node, the deployment has been done for collecting data from human body. The sensor nodes and sink node are placed on human body in equidistance from the entire deployed sensor nodes. After placing all the nodes at different locations on the body parts, distance connecting to the various nodes has been evaluated by the distance computation formula:

$$d = \sqrt{(x2 - x1)^2 + (y2 - y1)^2} \tag{1}$$

wherein d is the minimum distance between the nodes. For energy computation in data transmission, the distance between various nodes is calculated using Eq. (1).

C. ***Cluster Head Selection***

Dividing cluster has been done after the arrangement of sensor nodes in the body environment. Based on the density of source of node, clusters are separated in the network. The intervals between different cluster members are calculated, and energy is obtained for a particular after the cluster division. Node cluster head is calculated based on the utmost energy. The computed transmission energy equation is as given in Eq. (2):

$$\mathrm{ETX} = (\mathrm{ETX} + \mathrm{EDA}) * S * d^4 \tag{2}$$

Here, ETX indicates the energy transmission, EDA indicates energy available for data transmission, S indicates source, and d indicates destination.

The receiving energy equation is calculated using Eq. (3):

$$\mathrm{ERX} = (\mathrm{ERX} + \mathrm{EDA}) * S \tag{3}$$

Here, ERX represents the receiving energy.

The energy which is obtainable for a particular node is the amount of energy available to the node after the transmission of information from collected sensor nodes to base station, which is called as residual energy. The residual energy has been computed using Eq. (4):

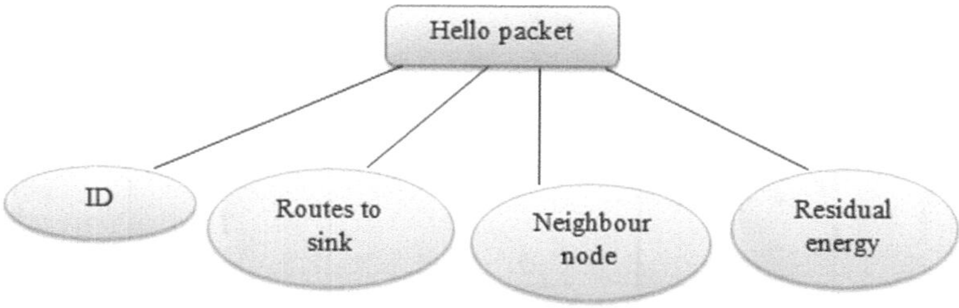

Fig. 3 The format of HELLO packet

$$\text{Residual Energy} = E - (\text{ETX} + \text{EDA}) * S * d^4 \quad (4)$$

where E indicates the total energy. Based on the mean distance and residual energy, the cluster head is selected. Cluster head node is the node that consists of more energy and less mean distance from all the nodes within the cluster. After the selection of cluster head, a "HELLO" message is transmitted by the cluster node along with its id, position, and neighbor nodes.

D. ***Start-Up Phase***

The sink transmits a "HELLO" data packet to every node in this phase. The format of the HELLO packet is given in Fig. 3. Every node accepts this data packet and saves the location of the sink. After that every node transmits a packet along with their corresponding ID, location, and the significance of the residual energy. All the nodes are rationalized with the position of neighboring nodes, location of the sink, and possible routes to the sink in the way [10].

E. ***Routing Phase***

After the assortment of cluster head, the information transmission from the source to routes has been started. In this procedure, these nodes sense information from the human body nodes to broadcast the data. After sensing the data, route is identified for transmitting the sensed information. Clusters are connected to the nodes to broadcast the sensed information to select a cluster head node, and in turn, cluster head measures the distance from sink node. TDMA approach is utilized for data transmission. This procedure that is used for broadcasting the information provides time period for all the nodes that are accessible in the network which is transmitted to the base station easily without any smashes.

F. ***Queuing Theory: Proposed Scenario***

The proposed scenario of queuing theory is a patient-centered WBAN used within the hospital site. ZigBee module is given to the patients as full functional device (FFD) mode. PAN coordinator is positioned in the vicinity; it is a committed ZigBee node which will always be present in the room and also in corridor to make easy for flawless communication even when the patient is moving in the hospital. With the assists of a bridging device, this PAN coordinator will transmit the information to the Wi-Fi router. Wi-Fi router (access point)

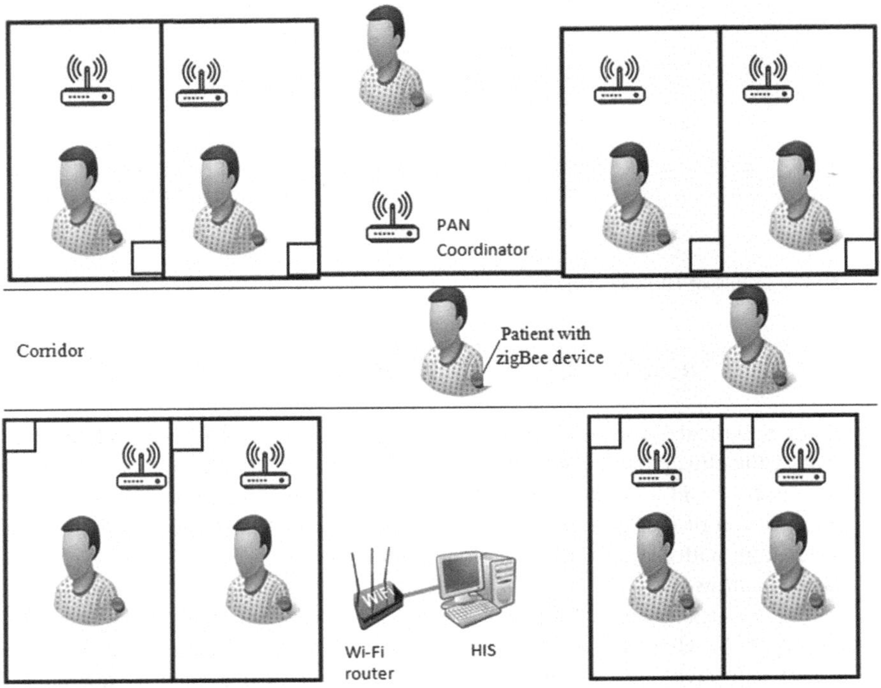

Fig. 4 Proposed scenario using queuing models

sends the data to the hospital information system using the Ethernet. The proposed queuing model is depicted in Fig. 4.

5 Modeling of WBAN Using Queuing Theory

Using heterogeneous network, the patient's vital parameter packets are sent to the HIS and are considered in the scenario. The PAN coordinator is hypothesized to be the data generator for the system, and the Wi-Fi router is treated as server to obtain the simple calculation. All packets reached the HIS, after getting the service from the access point. To follow the Poisson distribution and the service rate that follows general distribution, the arrival rate of the important parameter packets is assumed. At the Wi-Fi router, the special packets enter in the queue and transmit to HIS after receiving service. Single-hop transmission is assumed by the scenario, and the buffer capacity is hypothesized to be too maximizing than needed for the packet arrival rate and packet size. If the HIS is receiving the packet, it is saved for next processing. The vital parameter screening is done continuously for the patients. The intensive care unit (ICU) and general ward facility are considered by the hospital. Using ZigBee and Wi-Fi, every existing is understood to have a WBAN system for monitoring the vital parameters. These devices can be active at 24-h activation. Wi-Fi router for normal and emergency vital data packets is used by the priority queuing. Priority is given to the crisis vital data packets and service discipline is regarded as first in, first out (FIFO). The data may transmit at large quantity.

The objective of this study is to decide the following actions:

1. At Wi-Fi router, the average number of vital data packets is waiting in the queue (5):

$$E[NQ] = \sum_{n=1}^{\infty} (n-1)P(N = \mathrm{n}) = \rho^2/(1-\rho) \tag{5}$$

2. The crucial data packets in the queue, the average waiting time has

$$E[WQ] = E[T] - 1/\mu \tag{6}$$

   where $E[T] = E[N]$ / and $E[N] = \rho / (1 - \rho)$ (Little's law)
3. The server percentage utilization (Wi-Fi router) in the system (7):

$$P(N = n) = \rho^* n(1 - \rho) \tag{7}$$

   (a) In queue, it indicates fraction of time spent with no packet.
   (b) Utilization factor $= 1 - \mathrm{P(N = 0)} = \rho$ [10].

4. To facilitate efficient quality of service, a number of servers are desired.
   Here:

   λ: vital data packet arrival rate,
   μ: vital data packets service rate,
   P: /μ; < μ.
   N: total number of data packets in system.
   T: system packet time.
   NQ: data packets system in queue.
   WQ: queue waiting time.

6 Results And Discussion

In order to examine the effect of M/G/1 and M/G/N models on the proposed scenario, the simulations are conducted using MATLAB 7.2 on PC with 2.2 GHz Core i7 and 8GB of RAM. To forward the data from body sensors to hospital server, diverse vivacity effective steering plans are utilized. Behind the routing phase, data broadcasting has been completed. In the proposed scenario, packet transmit and network lifetime is computed for performance valuation for different energy levels. Network lifetime deals about network performance details. Since network lifetime increases, WBAN cost management decreases.

The spotted data of patient dependably got to curative pro for further examination, which is important. At all point, sink hub that leaves from broadcast range of hubs makes use of a hand-off hub which collected data from sensor hubs. In revolutionary convention, at whatever point it quiet shifts his hands, the remote connection of sink with sensor hubs disconnects. In the wireless sensor network, the vivacity consumption is a basic issue. To increase the throughput and reduce energy consumption, a new method is proposed. In our results we were able to remove these barriers all these issues by queuing theorem of different routing protocols in the WBANs which mainly focuses on link competence of networks that had high crash on network performance. The proposed method provides reliable and power-efficient routing protocol for WBANs. Based on proposed cost function, we are implementing WBAN routing protocol. WBAN with any nodes will be implemented to carry out these simulations. To initiate this simulation, 0.01 kilometer square has been considered as sensor field dimension. Then the receiving power, transmission power, and electronics power of every node in the simulated area will be calculated. The node will be simulated to die when the trx, txc, and elx power is low. The remaining alive nodes will be found out and the cluster node is a high energy node.

A. ***Performance Parameters***

Some of key parameters considered for performance of WBAN are given below:

1. Network lifetime: Total network process time.
2. Stability period: The time taken by the network operation until the last node dead is called stability period. The time span after the last node dead is called unstable period.
3. Throughput: The cumulative number of successful packets transmission.
4. in an established sink.
5. Residual energy: The energy expenditure of nodes per round is called as residual energy. It is used to monitor the energy utilization of network.
6. Path loss: The dissimilarity between the transmitted power and received power is called as path loss which is calculated in decibels (dB).

Existing system and proposed system are compared for total number of dead nodes as denoted in the following graph. The dead nodes are characterized by using this graph. The nodes whose energy is zero are called as dead nodes. The BEC protocol is used to represent by the red line, and the dynamic clustering is used to represent by the blue line as shown in Fig. 5. Since we used queue-based cluster selection, the energy consumption for data transmission is reduced to extend the lifetime of the network. The nodes which have sent and received the information effectively during transmission and whose level of energy is not zero are called alive nodes. Similarly existing system and proposed system are compared for total number of alive nodes as shown in Fig. 6. This graph is used to represent the proposed method outperforming than existing BEC model.

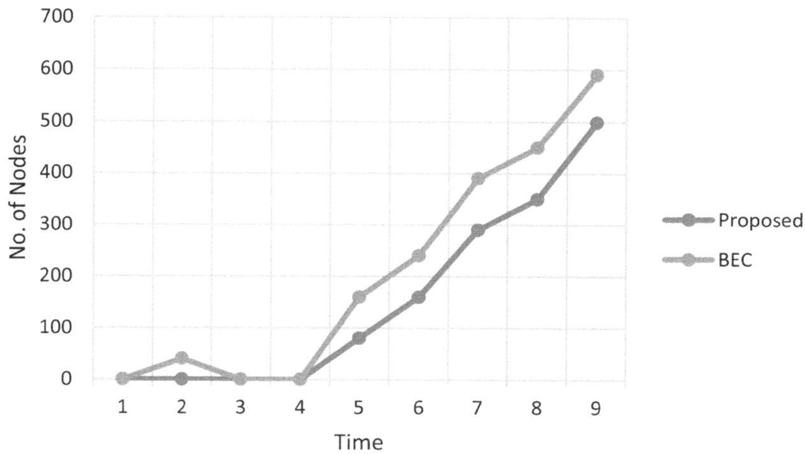

Fig. 5 Dead nodes representation

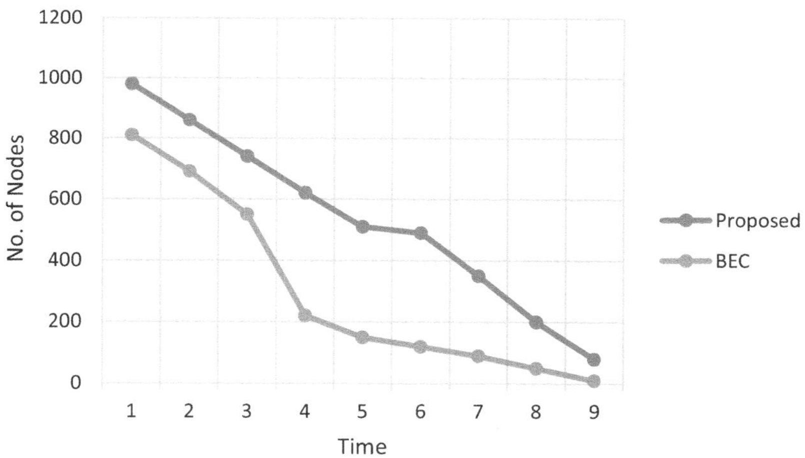

Fig. 6 Alive nodes representation

Based on total number of packets sent to base station, proposed method and existing methods were compared and denoted in Fig. 7. It determines how many packets are received by base station successfully. It drastically increased because of the better cluster selection using our queuing model. The BEC protocol is represented by red line and dynamic clustering is represented by blue line. Moreover, for M/G/1 and M/G/N queuing models, the simulated and mathematical analysis results are placed in this slot. The results of simulation study in form of a graph between average numbers of packets are shows in Fig. 8.

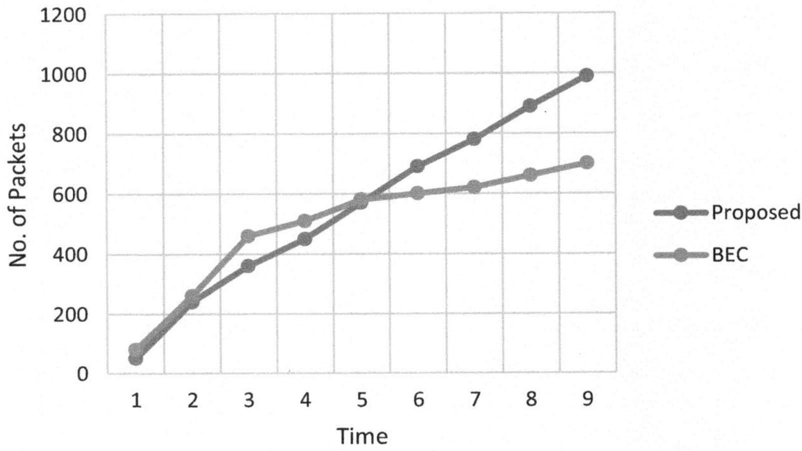

Fig. 7 Packet to BS representation

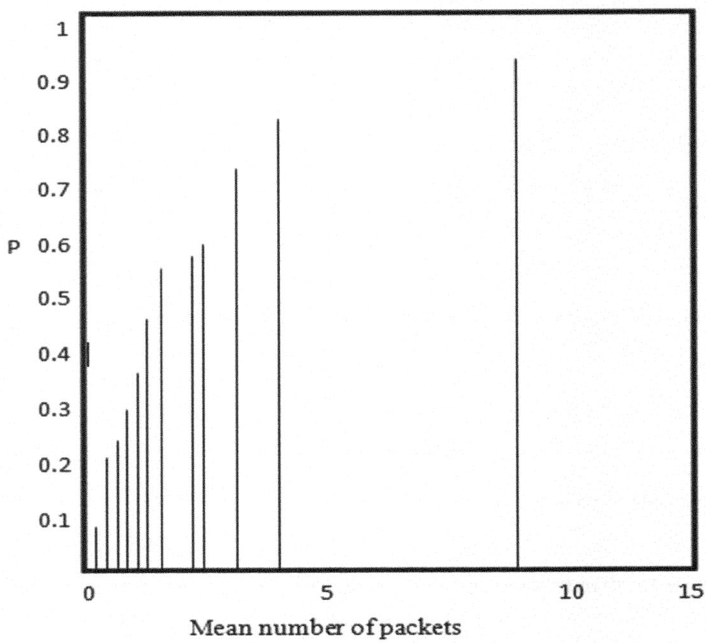

Fig. 8 Simulated results for M/G/1 queuing model for mean of packets in system

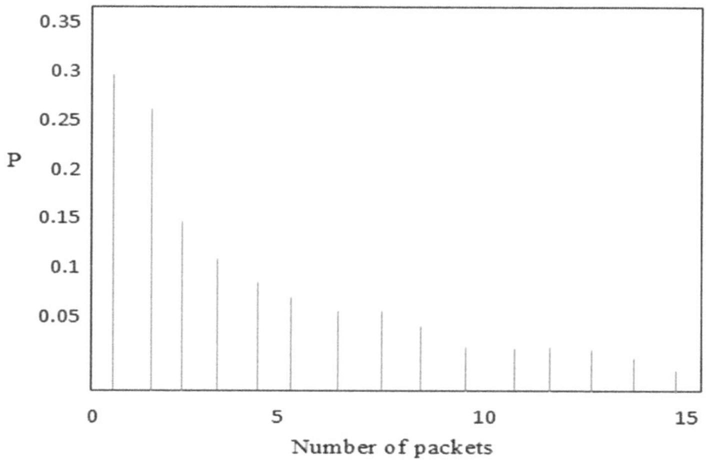

Fig. 9 Simulated results for M/G/N queuing model for mean number of packets in system

In the system with chances of occurrence at any known time in the system in the M/G/1 queuing model, Fig. 9 shows the comparison graph for M/G/N queuing model when the number of servers was enlarged, N taking values as 2, 3, and 4. With the help of this revision, we examined so as to for superior bandwidth (BW), Wi-Fi router, and probability of having slighter average number of packets in the system raises if we are using more than one router, but the system utilization is very poor. On opposing, if slighter BW, Wi-Fi routers are used, there is important enhancement in system operation. The server utilization graph for the mentioned discussion is shown in Fig. 10.

M/G/1 and M/G/N models are analyzed by the comparative study shown in Table 2; we deduce that when the number of servers is more in M/G/N model, it provides a better result for all the quality of service parameters such as estimated number of packets in queue (Nq), estimated number of packets in system (N), and average waiting time in queue (Wq). But as the number of servers is going up from 1 to 4, the server operation at any given point of time is started losing; consequently we conclude that when the amount of emergency patients is high and we need to make the amount of servers high, we can apply the servers with lesser bandwidth and lesser radiating power. It will not only decrease the delay but also develops the server utilization and causes a lesser amount of harmful effects on patients due to radio frequency radiations. Moreover it conserves the power requirements of the system, which is categorized by the area of research.

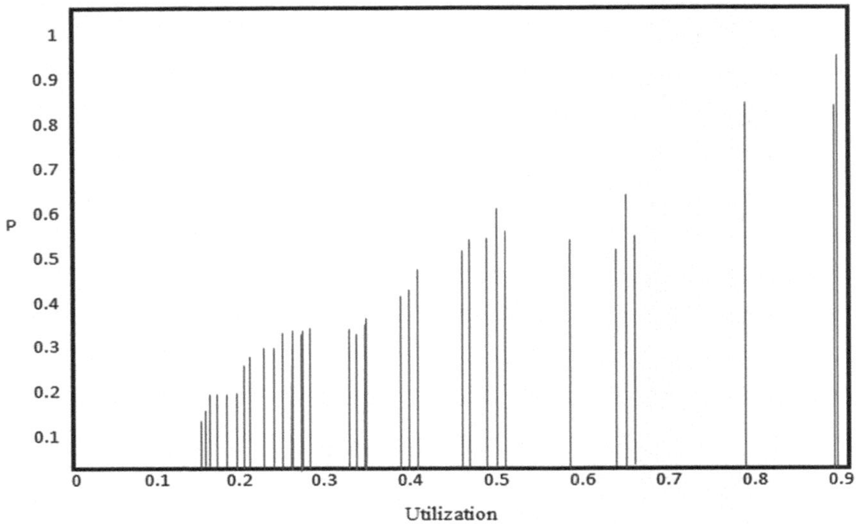

Fig. 10 Simulated results for M/G/1 queuing model for server utilization

Table 2 M/G/1 and M/G/N queuing models proposed scenario result

S. no	No. of servers	System utilization (%)	Expected no. of packets in queue (NQ)	Expected no. of packets in system (N)	Average queue waiting time (WQ)	Average system waiting time (W) (includes service)
1	4	45	0.105	1.905	0.043	0.793
2	3	60	0.532	2.332	0.221	0.971
3	2	90	7.673	9.473	3.197	3.947
4	1	100	8.0	10.01	5.236	4.212

7 Conclusion and Future Enhancement

Environment monitoring, habitant scanning, battlefield, farming field monitoring, and smart homes are monitored by wireless sensor networks with certain parameters in many applications. A subfield of WSN is WBAN. To transfer the data from human body sensors to hospital server, different energy-efficient routing methods are used. Patient information from body sensors should be reliably received by medical specialist for future analysis is significant. The proposed WBAN energy alert clustering-based routing protocol is used for sensing data from human body and sends this to base station for processing. Based on minimum distance and maximum energy from cluster head, the various clusters have been divided in the entire network. Best cluster head is chosen by the cluster members. After every round, the head selection is energetically changed in the cluster. In single round, based on dynamic cluster head selection, all nodes have a chance to play as a cluster head.

Cluster head sends out hello packet to every node which includes data about available route, ID, bandwidth, and energy available. For data transmission from base station, cluster head is responsible either directly or through any other cluster head accessible in the network. Finally, various types of parameters are obtained, i.e., throughput, residual energy, network lifetime, and network path loss. Moreover in this paper, we proposed the solution to the crisis handling in lack of human resources in the hospital management system. By organizing queuing with WBAN, we can save many lives by appropriate assistance and at the same time provide a sigh of break to the monitoring staff in panic situations.

The proposed holdup-tolerant WBAN with queuing method is mainly concerned with essential data information to take timely measures to improve the quality of service and decrease human resource intervention. This analysis can also aid in rating ruin management effectiveness of a hospital. We conclude that all the parameters of the proposed system give better results than other conventional methods. In the future, this work can be extended to optimize server utilization and average waiting time in queue parameter to improve the quality of service of the system and to reduce the total cost of the system. And this study can be used in cardiology and neurology hospitals where a patient usually comes with severe crisis and prone to sudden serious conditions which require a timely response. This work can be extended at next level using different performance metrics.

References

1. Adwan Alanazi (2016) Optimized node selection process for quality of service provisioning over wireless multimedia sensor networks. In: IEEE international conference on mobile and secure services, pp 1–5
2. Nadeem Javaid, Muhammad Moid Sahnadu, Mohsen Guizani, Muhammad Imran, Zahoor Ali Khan, Umar Qasim (2016) BEC: a novel routing protocol for balanced energy consumption in wireless body area networks. IEEE, pp 653–657
3. Ha I (2015) Technologies and research trends in wireless body area networks for healthcare: a systematic literature review. Int J Distrib Sens Netw 2015:14
4. Shah M (2015) Interference aware inverse EEDBR protocol for underwater WSNs. In: IEEE International conference on wireless communications and mobile computing, pp 739–744
5. Omer S (2015) An LQI based dual-channel routing protocol for wireless body area networks. In: IEEE international conference on telecommunication networks and applications, pp 320–325
6. Chen X (2015) Design and implementation of a body monitoring baseband system for human body communication. In: IEEE international conference on ASIC, pp 1–4
7. Gherbi C (2015) Distributed energy efficient adaptive clustering protocol for large scale wireless sensor networks. In: IEEE international conference on programming and systems, pp 1–7
8. Goel N (2015) Smart grid networks: a state of the art review. In: IEEE international conference on signal processing and communication, pp 122–126
9. Karthiga I (2015) A study on routing protocols in WBAN and its suitability for m-Health applications. In: IEEE international conference on communications and signal processing, pp 1064–1069

10. Sharma K, Singla S Wireless body area network in health care applications. Int J Sci Res (IJSR). ISSN (Online): 2319–7064
11. Dangi KG, Panda SP (2016) Performance analysis of patient monitoring WBAN using queueing models. Int J Appl Eng Res 11(9):6671–6666. ISSN 0973-4562
12. Ivanov S, Balasubramaniam S, Botvich D, Foley C (2012) Virtual groups for patient WBAN monitoring in medical environments. IEEE Trans Biomed Eng 59:3238–3246
13. Chen B, Pompili D (2010) Transmission of patient vital signs using wireless body area networks. Mobile Netw Appl 16(6):663–682. https://doi.org/10.1007/s11036-010-0253
14. Nidhya R, Karthik S (2019) Security and privacy issues in remote healthcare systems using wireless body area networks. In: Body area network challenges and solutions, EAI/ Springer innovations in communication and computing. Springer, Cham, pp 37–53
15. Ameen MA, Ullah N, Chowdhury MS, Islam SR, Kwak K (2012) A power efficient MAC protocol for wireless body area networks. J Wirel Commun Netw 33:17
16. Ragesh GK, Baskaran K (2012) An overview of applications, standards and challenges in futuristic wireless body area networks. Int J Comput Sci Issues 9:180–186
17. Jovanov E, Milenkovic A (2011) Body area networks for ubiquitous healthcare applications: opportunities and challenges. J Med Syst 35:1245–1254
18. Nidhya R et al (2022) Remodeled chaotic compressive sensing scheme for secure and energy-efficient data forwarding in body-to-body network. Comput Electr Eng 97:107633
19. Raghunathan V et al (2004) Energy efficient wireless packet scheduling and fair queueing. ACM Trans Embed Comput Syst 3:3–23
20. Tickoo O, Sikdar B (2004) A queueing model for finite load IEEE 802.11 random access MAC. In: IEEE international conference on ICC, France
21. Nageswari D, Maheswar R, Kanagachidambaresan GR (2018) Performance analysis of cluster based homogeneous sensor network using energy efficient N-policy (EENP) model. Clust Comput 22:12243–12250
22. Malathy S, Rastogi R, Maheswar R, Kanagachidambaresan GR, Sundararajan TVP, Vigneswaran D (2019) A novel energy-efficient framework (NEEF) for the wireless body sensor network. J Supercomput 76:6010–6025
23. Jayarajan P, Kanagachidambaresan GR, Sundararajan TVP, Sakthipandi K, Maheswar R, Karthikeyan A (2018) An energy aware buffer management (EABM) routing protocol for WSN. J Supercomput 76:4543–4555
24. Jayarajan P, Maheswar R, Kanagachidambaresan GR (2017) Modified energy minimization scheme using queue threshold based on priority queueing model. Clust Comput 22:12111–12118
25. Thirumoorthy P, Kalyanasundaram P, Maheswar R, Jayarajan P, Kanagachidambaresan GR, Amiri IS (2019) Time-critical energy minimization protocol using PQM (TCEM-PQM) for wireless body sensor network. J Supercomput 76:5862–5872
26. Jayarajan P, Maheswar R, Kanagachidambaresan GR, Sivasankaran V, Balaji M, Jagannath Das (2018) Performance evaluation of fault nodes using queue threshold based on N-Policy priority queueing model. IEEE – ICCCNT 2018, IISc, India
27. Jayarajan P, Maheswar R, Sivasankaran V, Vigneswaran D, Udaiyakumar R (2018) Performance analysis of contention based priority queuing model using N-Policy model for cluster based sensor networks. IEEE – ICCSP 2018, India
28. Ashraf Darwish, Kanagachidambaresan GR, Maheswar R, Lakthari a KI, Mahima V (2017) Buffer capacity based node life time estimation in wireless sensor network. In: The Eighth IEEE international conference on computing, communication and networking technologies (ICCCNT), IIT Delhi, 3–5 July
29. Bisnik N, Abouzeid A (2006) Queuing network models for delay analysis of multihop wireless ad hoc networks. In: International wireless communications and mobile computing conference (IWCMC 2006) Vancouver, IEEE Communication Society and ACM, July

30. Mohsin Iftikhar, Hossain MA (2011) An analytical G/M/1 queueing model for providing guaranteed QoS in multimedia surveillance. In: Proceedings of IEEE ICME 2011, Spain
31. Mohsin Iftikhar, Nada Al Elaiwi, Mehmet Sabih Aksoy (2014) Performance analysis of priority queuing model for low power wireless body area networks (WBANs). In: The 2nd international workshop on communications and sensor networks (ComSense-2014), pp 518–525

R. Nidhya is presently working as Professor in the Department of Computer Science and Engineering, Madanapalle Institute of Technology & Science, affiliated to Jawaharlal Nehru Technical University, Anantapuramu, India. She received the M. Tech and Ph.D. degrees from Anna University, Chennai. Her research interests include wireless body area networks, network security, and machine learning. She has published many research papers in refereed international journals and IEEE- and Springer-associated conferences. She is an active member of professional societies such as CSI, ISTE, SAISE, and ISRD.

Pavithra D is currently working as Assistant Professor in Dr. NGP Institute of Technology, Coimbatore, Tamil Nadu, India. She received a bachelor's degree B.E. (computer hardware and software engineering) from Avinashilingam Deemed University, Coimbatore, in 2002. Then she completed her M.E. (computer science) in Dr. Mahalingam College of Engineering and Technology, Pollachi, under Anna University, Coimbatore, in 2011. She completed her Ph.D. under Anna University Chennai in 2021. Her current research interests include autism, machine learning, and deep learning.

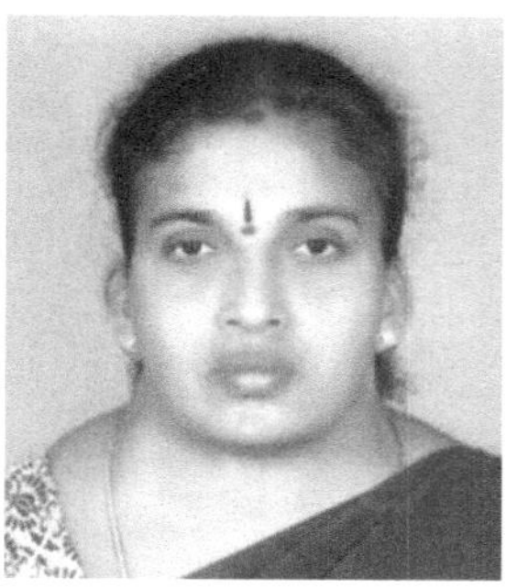

R. Kalpana received B.E. and M.E. degrees from Madras University and Sathyabama University in 1991 and 2006, respectively, and Ph.D. degree from Anna University Chennai, in 2013, all in computer science and engineering. She is working as a Professor in Computer Science and Engineering Department in Madanapalle Institute of Technology & Science, Madanapalle, Andhra Pradesh, India. Her research interests include network security, wireless networks, ad hoc networks, databases, and artificial intelligence.

M. Kathirvelu has completed B.E. at Government College of Engineering, Salem, in the year 1998 and M.E. from Kongu Engineering College, Perundurai, in the year 2004 under Anna University, Chennai, and secured university eighth rank in M.E. Degree Examination. He has received Ph.D. degree from Anna University, Chennai, in the area of low-power VLSI design during the year 2013 and presently works in the area of energy optimization. He has about more than 22 years of teaching and research experience and currently works as Head of ECE Department, KPR Institute of Engineering and Technology, Coimbatore, India.

P. Jayarajan has completed his B.E. (EEE) from Madurai Kamaraj University in the year 2004 and M.E. (applied electronics) from Anna University in the year 2008 and been pursuing Ph. D. in the field of wireless sensor network under Anna University. He has about 14 years of teaching experience and presently works as an Associate Professor in the Electronics and Communication Engineering Department, Sri Krishna College of Technology, Coimbatore. He has published 30 papers at international journals and conferences. His research interest includes wireless sensor network, modeling, and simulation.

Mechanical Energy Harvesting Scheme, Implementation Aspects, and Applications

Prem Prakash Singh, Anil Kumar Singh, Shivam Nigam, and Mahesh Kumar Singh

1 Introduction

Since its debut, electronics have shrunk and gained capability. Small, portable, yet powerful devices let users listen music and speak wirelessly through Bluetooth-based mobile device communication and wireless networking [1]. Wireless devices have advantages over wired alternatives. Adaptability, convenience of deployment, and the capacity to position sensors in unfavorable places are key [2]. As these gadgets get smaller, they're worn as accessories on watches, eyewear, apparel, and other products. In addition, our daily life is boosted by micro devices for application ranging from structural health monitoring, medical implants, industrial process, self-powered sensors, mobile electronics, and wireless sensor networks [3, 4]. Current microelectronics require external power for these devices. To supply the power to these micro devices, we have on-board storage components such as batteries, capacitors, and super-capacitors. The storage components possess less energy which results in short run-time for micro devices [5]. In addition to this, in distributed sensor networks, upgrading batteries on a constant schedule is a time-consuming and costly operation [6]. Since extended device life is crucial for the functioning of the device, therefore the crucial continuing strategy ought to be unconstrained of the meager power supply present throughout the operation or functioning of these devices. To address this issue, it is required to look on alternative energy solutions of power supply technologies.

P. P. Singh (✉) · A. K. Singh
Ewing Christian College, University of Allahabad, Prayagraj, India

S. Nigam
Colorado School of Mines, Golden, CO, USA

M. K. Singh
Indian Institute of Science, Bangalore, India

A. Nella et al. (eds.), *Energy Harvesting Trends for Low Power Compact Electronic Devices*, EAI/Springer Innovations in Communication and Computing,
https://doi.org/10.1007/978-3-031-35965-1_10

In this regard, energy harvesting has gained an attention as a substitute solution for low-power electronic devices [7]. In energy harvesting scheme, the energy is drawn and converted from the ambient environment energy into electrical energy. By supplying limitless clean energy, this technique not only minimizes the volume of non-ecological waste generated by battery sources but also increases the time span of electronic devices [8]. The energy source, the harvesting device, and load comprise a conventional energy harvesting system. The ambient energy or externally deployed source is used as the energy source for scavenging electrical energy. The device that will transform the external energy into electrical energy is the harvesting mechanism.

2 Energy Source

There are numerous diverse sources of energy that may be harvested, including light sources, RF sources, thermal energy sources, and mechanical energy sources. Classification of energy harvesting sources with the respective transduction methods is shown in Fig. 1.

Light Source
The energy harvested is inadequate to be used as the only power source for sensor networks since the energy obtained from the solar intensity is widely diverse and depends on day and night, weather, temperature, topographical position, and the interior or exterior surroundings [9]. Light energy is harvested through solar cells [10] and photovoltaic (PV) cells [11]. Because solar cells provide good power

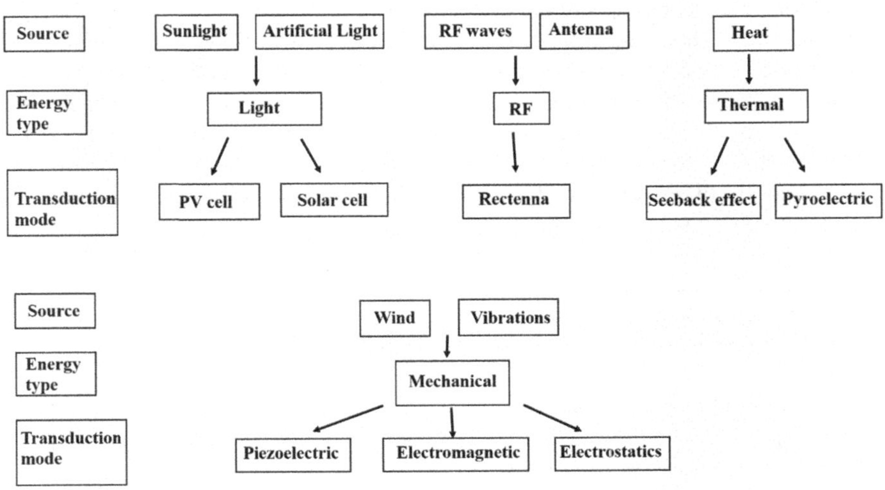

Fig. 1 Energy harvesting sources with the respective transduction methods [8]

density in direct sunlight, the highest rated type of source of light for outdoor harvesting is solar energy. However, power density is constrained, and artificial illumination levels are often below 1000 lux in dim ambient light circumstances [9].

Interior harvesting using manufactured lighting like compact fluorescent lamp (CFL) and light-emitting diode (LED) has gained popularity in recent years [12]. The desire to add to the growing knowledge on how to efficiently harvest ambient light energy in office spaces and homes for powering a variety of electronic devices, particularly Internet-of-Things smart sensors, has been the primary motivation for research development. Photovoltaic (PV) cells is perfect pick for harvesting interior external light, even in poor intensity situations, and utilizing it to power micro devices. PV cells are made up of different cell technologies such as amorphous silicon, crystalline silicon, and dye-sensitized and III-V semiconductors and have been investigated under poor light situation for interior light harvesting [13]. In fact, this technology has sparked considerable demand for the Internet of Things (IoT), smart appliances, and intelligent homes [14].

2.1 Radio Frequency Source

Radio frequency energy is copious and limitless in space. High electromagnetic field sources, such as satellites, wireless networks, cell phone towers, and digital multimedia transmission, all emit radio frequency (RF) waves. Radio waves can be received indoors as well as outdoors, in both agricultural and city areas, and at all times of the day. Although RF waves in the environment have a low power density, a necessary external source should be introduced for effective energy transmission. A rectifying antenna composes of a receiver module, a matched circuit, and a rectifying circuit, and it can be used to convert energy from radio frequency (RF) waves into direct current (DC) power [8]. The DC power that can be generated is determined by the input RF power, the antenna used, and the operating frequency [15]. The performance of RF energy harvesting is impacted by real-world constraints such as impedance mismatch, government laws, and harvested-power distribution.

This harvesting technique has recently been identified as a viable way to power wireless devices. In comparison to earlier technologies like the wireless powered network (WPN), the radio frequency harvesting technique has made it possible for wireless nodes to recharge using RF signals [16]. Simultaneous wireless information and power transfer (SWIPT) is a technology that combines wireless powering and transfer of information to do RF harvesting [17]. Implantable electronic devices and radio frequency identification (RFID) tags both employ the RF power harvesting approach. In RFID tags, the sensing device transmits a radio frequency wirelessly to a harvesting system that only provides sufficient power to transmit out identifying data particular to the target object. The majority of wireless sensor nodes in use utilize a little amount of power while sleeping and a large amount while operating [15]. A RF-DC converter circuit functioning at 906 MHz is used in the efficient far-field energy harvesting [18] to obtain power of up to 5.5 W. In their study of ambient RF energy harvesting, Bouchouicha et al. [19] used two systems to recover the RF energy: broad band without matching and narrow band. Further, rectifier topologies and ideal power transfer techniques were devised to enhance the sensitivity and performance of energy conversion from antenna to rectifier.

2.2 Thermal Energy Source

Heat continues to be a nearly universal and accessible ambient energy source in thermal energy harvesting, although it is frequently lost as low-grade waste heat. Thermal energy is converted to electrical energy using thermoelectric materials, pyroelectric materials, and ferroelectric materials. Attachable medical gadgets, electronic timepieces, self-driven thermal sensors, and cell phone technology are examples of prospective thermal energy harvesting uses [1]. A thermoelectric energy harvesting system is composed of thermoelectric generators (TEG), heat source, heat sinks, direct current-direct current (DC) converter, and DC load. Utilizing the Seebeck effect, thermoelectric generators generate electricity. In the Seebeck effect, an electric potential is produced by the temperature difference at the junctions of two thermoelectric materials. In pyroelectric energy harvesting, electrical heat is harvested from time-dependent temperature fluctuations.

Thermoelectric generators are utilized to generate both low power and high power. Early thermoelectric micro generators only produced a few nW [20] and were primarily used in electronic systems, notably smart wearable gadgets that harvest body temperature to produce electricity [21]. However, this strategy has recently been paired with micro-combustion chambers to increase power rating to 1 W per thermocouple [22]. Various manufacturing operations including those that produce vehicle engines and steelworks make use of high power generation. When the temperature difference is 1 to 3 °C, commercial electronic wristwatches generate 1.5 W or more for thermal conversion [23]. For the purpose of generating power from body temperature, micro thermoelectric generators [24] are explored. It is stated to have a 40 W power at 3 V and a five-degree temperature variance. The power conversion efficiency is significantly affected by the temperature differential over the TEG system and the thermoelectric figure of merit [25]. The poor efficiency of thermal energy harvesting (5–6%) could be a significant hindrance to its widespread usage.

2.3 Mechanical Energy Source

Mechanical energy is harvested primarily from strain, shocks, deformations, and vibrations. The most frequent types of vibrations are mechanical and fluid vibration. Human activity, vehicle motion, machinery, high-pressure motors, and home goods all contribute to mechanical vibrations, whereas, wind, ocean waves, and water movement are all examples of fluid vibrations. Fluid vibration energy is high power energy with the advantages of high energy density, global release, and limited effect on the environment. However, it has drawbacks as well, like being significantly impacted by the monsoon, requiring high efficiency equipment, and being challenging to manage and maintain [26]. Energy for microelectronic systems and low-power wireless sensor networks can be obtained from mechanical vibration harvesting systems. If the vibrational frequency of the ambient source is comparable to the natural frequency of the acquisition device, the harvesting system performs well [27]. Common domestic items like refrigerator, washers, dryers, ovens, etc., heavy manufacturing machinery, mobile components like cars and aircraft, and constructions all use the appropriate vibrations which can be used for harvesting purpose [28]. Naturally, the frequency range of vibrations is wide and is below 100 Hz.

Piezoelectric, electromagnetic, electrostatic, and magnetostrictive are a few of the energy conversion techniques that have been used to convert mechanical energy. One may claim that low-power energy production from kinetic energy and environmental vibrations is a rather well-established field given the number of review articles that have already been published (Table 1).

Looking at the energy densities (refer to Table 1) and comparisons (refer to Table 2) of ambient or externally deployed energy sources such as light, RF, thermal, and mechanical, the mechanical energy source appears to be the most promising.

Table 1 Overview of power density and harvesting technology of different energy sources [29, 30]

Source	Power density	Harvesting tech.
Light	15 mW/cm^3	Photovoltaic Solar cell
RF	0.1 $\mu W/cm^2$–1 mW/cm^2	Antenna Radio waves
Thermal	30 $\mu W/cm^2$–10 mW/cm^2	Thermoelectric Pyroelectric
Mechanical	4 $\mu W/cm^2$–100 $\mu W/cm^2$	Piezoelectric Electromagnetic

Table 2 Comparison of different energy sources [29, 30]

Source	Advantages	Disadvantage
Light	High power density Developed	Occasionally available Light is required Costly
RF	Always available Embedded	Low density Efficiency decreases as source goes away
Thermal	High power density Embedded	Occasionally available Extra heat
Mechanical	High productivity Embedded	Occasionally available Material and physical restrictions

3 Transduction Mechanism for Mechanical Energy Harvesting

A transduction mechanism is required for mechanical energy harvesting to transform vibrational energy into electrical energy, and the generator will need a mechanical system to connect mechanical deformation to the transduction mechanism. The mechanical system's design will be completely determined by the properties of the mechanical vibrations and should attempt aim to maximize interaction between the vibrational energy source and the transduction mechanism. Piezoelectric, electrostatic, and electromagnetic are the three main types of transduction for mechanical energy harvesting.

By exploiting of the mechanical strain or relative displacement taking place within the system, the transduction mechanism can produce electricity on its own. The deformation of mechanical system and piezoelectric materials are used to provide the strain effect. Velocity and an electromagnetic transduction mechanism can be linked in the case of relative displacement, while electrostatic transduction can be linked to position. Different damping characteristics are displayed by each transduction mechanism, and this needs to be address while designing the generators [2].

3.1 Piezoelectric

Piezoelectricity can appear in two different ways depending on mechanical and electrical behavior of materials. When mechanical stress is introduced into piezoelectric materials, crystals become electrically polarized [31]. It is important to keep in mind that the material exhibits the inverse piezoelectric effect by applying electric field. Single-crystal materials, such as quartz and lithium niobate (LiNbO3); lead-based piezoceramics, such as lead zirconate titanate or PZT; lead-free piezoceramics; thin films, such as sputtered zinc oxide; and piezopolymers, such as polyvinylidene fluoride (PVDF), are just a few examples of the numerous types of piezoelectric (refer to Table 3). Conventional single crystals are used to make single-crystal piezoelectric materials; however these materials perform less effectively than piezoceramics in terms of piezoelectricity. Recent improvements in materials and processing techniques are too responsible for the current flood of piezoelectric products like gyroscopes and accelerometers. Since then, piezoelectric materials employed in vibrational energy harvesting mechanism have the advantage.

The coupling within piezoelectric material is given by [39].

Table 3 Comparison of piezoelectric material types [8]

Type	Characteristics	Examples
Single-crystal materials	Extraordinary piezoelectricity, primarily used in applications involving sensors and actuators	ZnO [32] PMN-PT [33]
Lead-based piezoceramics	Strong piezoelectric effect Small dielectric loss	PMN-PZT [34] PZT-5A [34] PZT/ZnO [35]
Lead-free piezoceramics	Lower transduction efficiency Nonhazardous	BaTiO3 [36] BNT-BKT [37]
Piezopolymers	Flexible and nonhazardous Minimal manufacturing cost Biocompatible	Piezo-MEMS PVDF polymers [38]

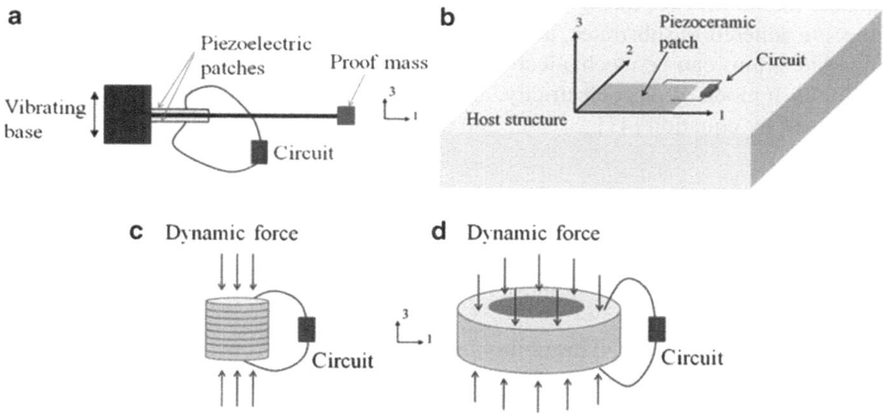

Fig. 2 Various piezoelectric energy harvester setups [40]

$$\begin{aligned} D_i &= \varepsilon_0 \varepsilon_{ij}^{\sigma} E_i + d_{iJ} \sigma_J \\ \delta_I &= S_{IJ}^{E} \sigma_J + d_{Ii} E_i \end{aligned} \tag{1}$$

where symbols have their usual meaning.

Piezoelectric materials often have anisotropic properties, which means that the material's characteristics rely on the polarization and force directions. Various piezoelectric energy harvester setups are depicted in Fig. 2. The device exploits the 31 mode of piezoelectricity when the electrode axis and the transverse strain axis are perpendicular to each other. In this mode, the 3 and 1 axes correspond to the polarization and strain axis, respectively. Bimorph cantilever with base vibration (Fig. 2(a)) and a piezoelectric patch to capture strain variations of the host structure

(Fig. 2(b)) are used for 31 mode of piezoelectricity. On the other hand, the piezoelectric device is reportedly functional in the 33 mode if the polarization and compressive mechanical strain axes coincide. Piezoelectric stacks (Fig. 2(c)) and monolithic cylindrical piezoelectric (Fig. 2(d)) are both employed for the 33 mode of piezoelectricity.

In a typical piezoelectric harvester, the transverse coefficient d_{31} is used as the piezoelectric coefficient if the voltage is obtained along the transversal axis of the piezoelectric layer and the deformation results in due to an applied load alongside longitudinal axis. This results in the following piezoelectric formulation:

$$\begin{aligned} D_3 &= \varepsilon_0 \varepsilon_{33}^{\sigma} E_3 + d_{31}\sigma_1 \\ \delta_1 &= S_{11}^{E}\sigma_1 + d_{31}E_3 \end{aligned} \tag{2}$$

The thickness of the piezoelectric element can be increased, or multilayer stacks can be used, to enhance the energy output obtained in the compressive mode. In contrast, the majority of applications do not make use of compressive loading as a suitable method for capturing vibrational energy. Typically, when the piezoelectric films are adhered to substrates, the elements are coupled in a vertical direction. The imparted strains can be mechanically enhanced [2] using a similar method.

For each mode of piezoelectricity, an electromechanical coupling coefficient can be defined by equation [41]

$$k_{ij}^2 = \frac{W_i^e}{W_j^m} \tag{3}$$

W_i^e is the stored electrical energy in the ith axis and W_j^m is the input mechanical energy in the jth axis.

This coefficient indicated the ability by which a material converts energy between electrical and mechanical forms. With a higher coupling coefficient, more mechanical energy is converted to electrical energy. Piezoelectricity can be found in both monocrystalline and polycrystalline ferroelectric ceramics. The number and diversity of thin film materials have increased. The choice of a piezoelectric material for a particular purpose is impacted by mechanical and dielectric losses, design flexibility, application frequency, and other factors. Crystalline and ceramic materials convert more efficiently ($k_{33} = 0.7$), but they are also the most sensitive and prone to aging. PVDF has greater mechanical resistance but a lower coupling coefficient ($k_{33} = 0.2$). Ceramics and polymers are combined in piezo-composites. Due to environmental constraints, their use is limited due to a small number of producers and the presence of lead in their composition. Many reviews have compared the benefits of the most common ones. A compressed piezo-stack's efficiency was researched by Goldfarb et al. [42], and efficiency was at its peak at manyfold lower than the resonance frequency.

Table 4 Piezoelectric constants of common piezoelectric material [43]

Property	PZT-5H	PZT-5A	$BaTiO_3$	PVDF	AIN
$d_{33}(10^{-12}\ CN^{-1})$	593	374	149	−33	5
$d_{31}\ (10^{-12}\ CN^{-1})$	−274	−171	78	23	−2
k_{33}	0.75	0.71	0.48	0.15	0.23
k_{31}	0.39	0.31	0.21	0.12	–
Relative permittivity (ε/εo)	3400	1700	1700	12	

Temperature, pressure, and aging all have an impact on piezoelectric characteristics. Producers typically describe the device's constants after a certain time duration. The variations in the properties of material have a tendency to be changing very little with changing time, and, as a result, the material properties stabilize with age. Due to the Curie point, temperature is a constraint for piezoceramics. Above this point, the material will no longer be piezoelectric and will instead become depolarized. This Curie temperature can also be lowered by applying stress. Table 4 lists the piezoelectric constants for five typical materials: polyvinylidene fluoride (PVDF), soft and hard lead zirconate titanate piezoceramics (PZT-5H and PZT-5A), barium titanate (BaTiO3), and AIN.

There are two types of potential converters with regard to mechanical structures. One uses inertial converters to convey the vibrations or displacement of a suspended mass. The energy gained in this condition will be impacted by this mass. There are inertial converters made of piezoelectric materials as well. In these devices, an accelerating mass deforms the piezoelectric material by high force or vibration. In the second instance, an outside force exerts pressure which is translated into elastic potential energy. This results in a compression that is then transformed to electrical energy by a piezoelectric material, also known as non-inertial converters. In this instance, the energy obtained is not affected by the converter's mass, but rather, generally, by the rate of deformation [44].

Materials

PZT is among the most studied and utilized piezoelectric materials; however due to lead toxicity, its use has been restricted by law. As a consequence, research has shifted to other substances with lesser transduction effectiveness, like barium titanate (BaTiO3). Subsequently, because they can sustain high strain, polymer piezoelectric composites are taken into consideration for applications requiring a large degree of bending. Vatansever et al. [45] noted that regardless of their poor capacitance and power density, they might produce additional voltage in comparison to piezoceramics under certain circumstances.

Monolithic PZT has been created by Sodano et al. [46] for generating 0.14 to 0.20 mW at 1 KHz. Piezofiber composite was employed by Mohammadi et al. [47], which has enhanced flexibility and produced 120 mW power from a 34 x 11 mm plate of 5.85 thickness. Funasaka et al. [48] also examined LiNbO3 plates' impact resistance. LiNbO3 was preferred due to its higher Q and k. PZT and LiNbO3 plates were 65% and 78% efficient under impact conditions. Hydraulic heel and sole systems linked to PZT stacks could provide parasitic power [49]. Hydraulics increase stack force while reducing stroke. Pondrom et al. [50] proposed using folded stacks to make PEHs. The prototype produced 1.3 mW at 140 Hz vibration and 1 g acceleration. MIT uses a 2 mm PVDF-electrode sheet with eight layers. A shoe's insole bending powered the d_{31} mode. This configuration delivered 1.3 mW at 250 kHz [51]. Feenstra et al. [50] recommended using a commercial PZT stack for energy collecting. They described replacing a backpack's buckle with a piezoelectric stack. The system produced an average power of 0.4 mW, which grew to 45.6 mW when a piezoelectric energy harvester based on polyvinylidene fluoride was added to backpack straps [52]. Kymissis et al. [53] developed a multilayer polyvinylidene fluoride insole for heel striking energy harvesting. The shoe heel deforms and resets transducers.

3.2 *Geometry of Transducers*

As shown in Fig. 3, piezoelectric transducers exist in various shapes like cantilever beam, circular diaphragm, cymbal type, and stack type.

The cantilever beam is fabricated by deposition of piezoelectric thin film on a non-piezoelectric substrate that is fixed at one end. Because of simple shape and capacity to produce the highest strain, it is the most frequently employed structure. The configuration is known as "unimorph" if there is only one layer of piezoelectric thin film deposited to the substrate and "bimorph" if there are two piezoelectric layers. Due to their ability to quadruple electrical energy production without increasing device volume, bimorph cantilevers are broadly widely employed in PEH.

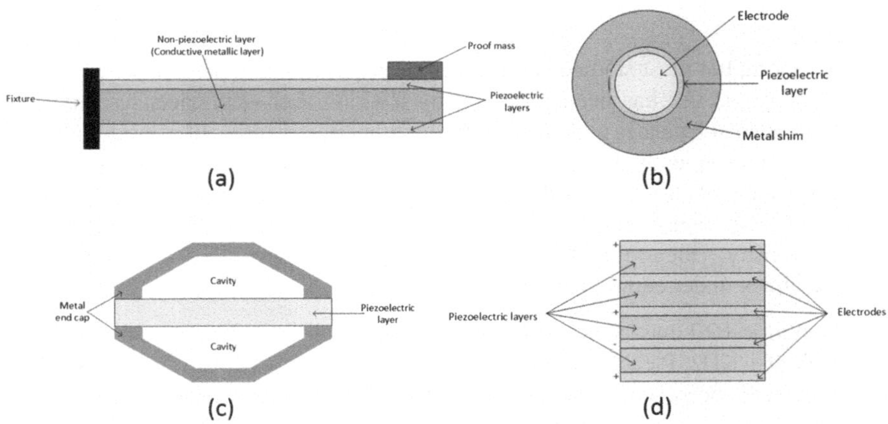

Fig. 3 Different shapes of piezoelectric transducers: (**a**) cantilever, (**b**) circular diaphragm, (**c**) cymbal type, (**d**) stack type [54]

In their design proposal for energy harvesting, Kymissis et al. [53] suggested using a unimorph cantilever curved beam. A 1.75 cm PZT bimorph cantilever was exploited by Rundy et al. [55] to capture energy from low-amplitude vibrations. In order to reduce the resonance frequency, a proof mass was attached to the cantilever's tip and able to generate power of 60 W. Kim et al. [56] harvested energy using a PZT cantilever beam and reached 2.23 W power at 243 Hz. The outcome demonstrates that the design of the electrodes affects energy harvesting using various piezoelectric modes. In order to execute energy harvesting depending on the effect of mass onto the cantilever tips, Renaud [57] combined two bimorph cantilevers as a harvesting system. At 10 Hz frequency, an output of 600 W was attained. A harvesting system of suspended mass on stiff piezoelectric bimorph cantilevers can operate at low frequency while a person moves, which was proposed by Zhang et al. [58]. Pillatsch et al. [59] have vibrated a matrix of piezoelectric bimorph beams by a cylindrical proof mass through magnetic attraction. For 2 Hz frequency and 2.7 ms^{-2} acceleration, the concept was found to provide a highest output of 2.1 mW.

The introduction of microscale energy harvesting system is a result of the advancement of microfabrication technology in recent years. Circular and square PVDF plates for blood pressure energy have been examined by Sohn et al. [60]. A circular diaphragm with a radius of 5.56 mm produces 0.61 W power for 9 m thick diaphragm and power of 0.03 W for 110 m thick diaphragm. In their evaluation of the MEMS-based energy harvesters that use piezoelectric thin films, Todaro et al. [61] highlighted that these harvesters are affordable and compact. They noted that expanding the frequency bandwidth and the amount of generated power are the two biggest obstacles to creating high-performance devices. A design composed of test mass coupled to one end of a 1 m thick piezoelectric MEMS cantilever is fabricated and studied by Ammar et al. [62]. Capacitive MEMS were examined by Wright et al. [63] for harvesting low-level vibrations that happen in typical home and office settings. Shen et al. [64] presented a PZT cantilever with silicon test mass for producing a 0.32 W of output power. A thick coating of PZT was used by Liu et al. [65] to create a MEMS-based generator that produced 3.98 mW of electricity.

1 μm thick aluminum nitride (AlN) and PZT deposited MEMS cantilever fabricated by dry etching of silicon-on-insulator wafer has been modeled. Simulations projected that AIN layer produced 100 nW and PZT layer produced 600 nW at 9 kHz resonant [66]. Another MEMS cantilever for energy harvesting is fabricated from a membrane-like structure. Membrane is a stack of PZT layer over thermally grown silicon oxide, buffered layer zirconium dioxide and silicon nitride [67].

Harvesting generators have been further miniaturized and improved in performance, thanks to the tremendous advancements in nanoscience and nanotechnology over the past few decades. The widely used piezoelectric component in nanoscale piezoelectric harvesting systems seems to have been ZnO. An elastic nanoscale piezoelectric harvesting device depending on the periodic elongation and release process of a zinc oxide nanowire placed on a flexible substrate and linked to metal electrodes was reported by Yang et al. [68]. The measured output voltage accelerated up to 50 mV, and the energy conversion efficiency was 6.8%. Numerous additional

Table 5 Comparison of different geometry of transducers [54]

Geometry of transducer	Advantage	Disadvantage
Cantilever beam	Low cost of fabrication Lesser resonant frequency Excellent mechanical Q factor	Unable to withstand a large applied load
Circular diaphragm	Suitable for operating in pressure method	By applying physical load, resist more in comparison to cantilever Higher resonant frequencies
Cymbal transducer	High output Sustains with strong applied load	Primarily suitable for uses demanding large vibrational amplitude
Stacked structures	Sustains with heavy load Appropriate for operating in pressure mode Large output from d_{33} mode	By applying physical load, resist more

piezoelectric materials, in addition to ZnO, also have uses at the nanoscale. An energy harvesting system that depends on vertically stacked PZT nanowires was created by Xu et al. [69]. The development of a piezoelectric harvesting scheme based on nanoribbon was reported by Dagdeviren et al. [70]. This design uses the normal systole and diastole motions of the respiratory system and demonstrated the potential to be used to power pacemakers. Additionally, an energy harvesting scheme using a piezoelectric micromachined ultrasonic transducer (pMUT) is presented [71].

Cymbal structures are effective for the micro energy harvesting since they can generate a significant in-plane strain when subjected to a transverse external force. Piezoelectric energy harvesting demonstrated excellent outcomes during periodic behavior of pre-stressed cymbal-type structures [72]. Due to the usage of piezoelectric materials in the d_{33} mode and having a large capacitance due to multiple stacked layers on top of each other, the stack-type piezoelectric transducer can generate a lot of electrical energy. Adhikari et al. [73] examined two situations, one with an inductor and the other without indicator in the circuit, and offered a probabilistic approach employing stack configuration as opposed to periodic actuation of cantilever at resonant condition (Table 5).

3.3 Electrostatics

In electrostatic transduction mechanism, energy is harvested by changing the capacitance of charged capacitor. The actuation of charged capacitors varies the distance between the plates, and in result, capacitance is changed. Variation in capacitance is dependent on variation in one of the two parameters, Q and V, while maintaining a fixed value for the other parameter. Therefore, charge-constrained and voltage-constrained techniques to harvest energy harvesting are frequently used approaches [74]. Voltage-constrained approaches typically give more energy than charge-constrained approaches. The energy from the charge-constrained technique approximate to the voltage-constrained technique by adding an external capacitor, parallel to the harvesting capacitor [75]. Electrostatic energy harvesting, in contrary to piezoelectric and electromagnetic energy harvesting, needs a DC voltage to counter charge the capacitor plates [76]. Despite being the primary flaw of electrostatic harvesters, this can be rectified by introducing a dielectric material which is permanently polarized [77].

Three different configurations of electrostatics generator are possible – in-plane overlap varying, in-plane vibrations, and out-of-plane vibrations of the capacitor plate with respect to other plates [76]. In-plane arrangements have two variable capacitors whose capacitances are 180° out of phase with each other. Both charge- and voltage-constrained approach can be used to operate the three possible configurations. In-plane vibrations deliver the maximum power output, with an optimum design yielding 100 W cm^3, according to Roundy et al. [78]. Out-of-plane vibration and in-plane overlap varying are the next highest. The production of the most power happens at very small dielectric gaps. Microfabrication techniques have been extensively used in the realization of MEMS-based energy harvesting systems which include both overlap varying and vibration mechanisms.

4 Applications

All proposed transduction mechanisms increase generated power. Vibration form determines the frequency spectrum of the harvesting mechanism, whereas vibration direction sets the harvesting orientation. The system designer can develop a system that is not capable of inducing resonance or extend the operating frequency range (WOF) to increase system frequency responsiveness. When the vibration frequency is known, WOF approaches are more applicable, whereas non-resonant systems are better at capturing low-frequency motions. Two-dimensional (2D) and three-dimensional (3D) harvesting approaches could be useful if the vibration is multidirectional.

Ultralow-power wireless sensor networks have industrial, medicinal, and commercial uses. This network power sensors in isolated or difficult-to-reach positions to alert about construction strains, environmental pollution, wild fires, and landslides. Many consumer and industrial products, such as wristwatches, electronic calculators, outdoor lighting systems, handheld power banks, toys, and spacecraft, use small solar cells. TE-CORE7 Thermal Energy Harvesting Modules from Micropelt transform excess heat for low-power devices. Piezoelectric transducers are electrically active while under pressure, which makes them suitable for use as mechanical sensors in energy harvesting mechanism that detect noisy bearing and aircraft wing turbulence. A pyroelectric generator was developed by Oak Ridge National

Laboratory that can generate electricity while cooling electronics, photovoltaic cells, computer systems, and large thermal waste producing equipment. The gadget uses a MEMS pyroelectric capacitor on an oscillating bimetal cantilever.

Based on 16-bit microcontroller, a self-powered security tool is designed [79], and this is appropriate for switching and distributes power between stations. In order to fuel the automatic electronic smart systems, mechanical energy harvesting techniques are used in the railway sector to transform the rotational motion of the rail axles into electric energy [80]. To provide powers to implanted hearts, blood pressure sensors, and heart and heartbeat sensors, Ali et al. [81] investigated piezoelectric energy harvesting from muscle contraction and elongation, blood flow, lung movements, and beating of hearts. For use of harvested bio-energy in body movement sensing, biodegradable contaminant decomposition, and microbial sterilization, Surmenev et al. [82] described fabrication of composite piezoelectric polymers. Mhetre et al. [83] reviewed micro energy harvesting for medicine delivery, dentistry uses, and body heat restoration.

5 Summary

The overview and comparison of harvesting technology of different energy sources are presented, and it is demonstrated that mechanical energy source is the most promising among all. The objective of this chapter is to present the most recent research developments in mechanical energy harvesting with an emphasis on various transduction mechanisms for mechanical energy harvesting like piezoelectric, electrostatic, and electromagnetic. It is demonstrated that the mechanical energy from the mechanical vibrations, mechanical strains, and mechanical deformations is harvested to produce electricity which is supplied to low-power smart systems. Various shape and geometry of harvesters are discussed in this chapter. It is shown that energy harvesters are miniaturized using MEMS technology which reduces the cost of production of harvesting system. Numerous research teams are examining potential application of self-powered wireless sensor systems in the field of medical engineering, bioengineering, and wearable technology.

References

1. Moll F et al (2005) Review of energy harvesting techniques and applications for microelectronics. Proc SPIE 5837, VLSI Circuits and Systems II (2005). https://doi.org/10.1117/12.613046
2. Beeby SP et al (2006) Meas Sci Technol 17:R175
3. Roundy S et al (2004) Energy scavenging for wireless sensor networks: with special focus on vibrations. Kluwer Academic, Boston, MA
4. Priya S et al (2009) Energy harvesting technologies. Springer, New York
5. Abdulkadir M et al (2013) Mechanical energy harvesting devices for low frequency applications: revisited. ARPN Journal of Engineering and Applied Sciences 8(7)

6. Glynne-Jones P, White NM (2001) Self-powered systems: a review of energy sources. Sens Rev 21(2):91–98
7. Chu F et al (2018) Hybrid energy harvesting from mechanical vibrations and magnetic field. Appl Phys Lett 113:013901
8. Tounsia F et al (2019) A review on design improvements and techniques for mechanical energy harvesting using piezoelectric and electromagnetic schemes. Energy Convers Manag 199: 111973
9. Aurel Chirap et al. (2014) A study on light energy harvesting from indoor environment 12th international conference on development and application systems, Suceava, Romania, may 15-17
10. Chen P et al (2019) Energy harvesting under dim-light condition with dye-sensitized and perovskite solar cells. Front Chem 7:209
11. Yan J et al (2022) Solar energy harvesting technologies for PV self-powered applications: A comprehensive review. Renew Energy 188:678–697
12. De Rossi F et al (2015) Characterization of photovoltaic devices for indoor light harvesting and customization of flexible dye solar cells to deliver superior efficiency under artificial lighting. Appl Energy 156:413–422
13. Reich NH et al (2011) Charge yield potential of indoor-operated solar cells incorporated into product integrated photovoltaic (PIPV). Renew Energy 36:642–647
14. Tudor MJ et al (2014) Review of the application of energy harvesting in buildings. Meas Sci Technol 25:012002
15. Zungeru AM et al (2012) Radio frequency energy harvesting and management for wireless sensor networks. arXiv:1208.4439
16. Schober R et al (2015) Energy-efficient resource allocation for wireless powered communication networks. IEEE Trans Wirel Commun. https://doi.org/10.1109/TWC.2015.2502590
17. Ding Z et al (2015) Application of smart antenna technologies in simultaneous wireless information and power transfer. IEEE Commun Mag 53(4):86–93. https://doi.org/10.1109/MCOM.2015.7081080
18. Fiez T et al (2008) Efficient far-field radio frequency energy harvesting for passively powered sensor networks. IEEE J Solid State Circuits 43(5)
19. Ventura L, et al (2010) Ambient RF energy harvesting. International conference on renewable energies and power quality (ICREPQ'10) Granada (Spain), 23th - 25th March, 2010
20. Rowe D et al (1991) Low cost miniature thermoelectric generator. Electron Lett:27
21. Jeffrey Snyder G (2008) Small thermoelectric generators. Electrochem Soc Interface 17:54
22. Washabaugh PD, et al. (2001) An integrated combustor-thermoelectric micro power generator. Transducers '01 Eurosensors XV, 11th International Conference on Solid-State Sensors and Actuators, Munich, Germany, June 10–14, 2001
23. Seiko Instruments Inc. http://www.sii.co.jp/info/eg/thermic main.html
24. Applied Digital. http://www.adsx.com/prodservpart/thermolife.html
25. Chen WH et al (2012) Design of heat sink for improving the performance of thermoelectric generator using two-stage optimization. Energy 39:236–245
26. Yuhao W et al (2021) A review on vibration energy harvesting. E3S Web of Conferences 245: 01041
27. Harne RL, Wang KW (2013) A review of the recent research on vibration energy harvesting via bi-stable systems. Smart Mater Struct 22(2)
28. Roundy S, Wright PK, Rabaye J (2003) A study of low level vibrations as a power source for wireless sensor nodes. Comput Commun 26:1131–1144
29. Mertens R et al (2009) Micropower energy harvesting. Solid State Electron 53:684–693
30. Akhtar F, Rehmani MH (2015) Energy replenishment using renewable and traditional energy resources for sustainable wireless sensor networks: a review. Renew Sust Energ Rev 45:769–784
31. Elvin N, Erturk A (2013) Introduction and methods of mechanical energy harvesting in advances in energy harvesting methods. Springer, Berlin, Germany, pp 3–9

32. Rajagopalan P, Singh V, Palani I (2018) Enhancement of ZnO based flexible nano generators via sol gel technique for sensing and energy harvesting applications. Nanotechnology
33. Schmidt OG et al (2017) Scalable single crystalline PMN-PT nanobelts sculpted from bulk for energy harvesting. Nano Energy 31:239–246
34. Friswell MI (2011) Experimental and analytical parametric study of single-crystal unimorph beams for vibration energy harvesting. IEEE Trans Ultrason Ferroelectr Freq Control 58(7): 1508–1520
35. Li H et al (2011) Preparation and characterization of PZT thick film enhanced by ZnO nanowhiskers for MEMS piezoelectric generators. Progr Nat Sci Mater Int 21(2):159–163
36. Park K et al (2017) Lead-free BaTiO 3 nanowire arrays based piezoelectric energy harvester. MRS Adv 2(56):3415–3420
37. Lee G et al (2016) Optimized piezoelectric and structural properties of (bi, Na) TiO3-(bi, K) TiO3 ceramics for energy harvester applications. Ceram Int 42(13):14355–14363
38. Farinholt K, at al. (2007) Energy harvesting from a backpack instrumented with piezoelectric shoulder straps. Smart Mater Struct 16(5):1810
39. Evoy S et al (2014) A review of piezoelectric polymers as functional materials for electromechanical transducers. Smart Mater Struct 23(3)
40. Elvin N, Erturk A (2013) Introduction and methods of mechanical energy harvesting. In: A. (ed) Advances in energy harvesting methods. Springer, New York, NY. https://doi.org/10.1007/978-1-4614-5705-3_1
41. Schmidt MA et al (2001) A combustion-based MEMS thermoelectric power generator. In: Proc. 11th Int. Conf. On solid-state sensors and actuators, transducers 01, Munich, Germany, pp 1A3–1A102
42. Goldfarb M, Jones LD (1999) On the efficiency of electric power generation with piezoelectric ceramic. Trans ASME J Dyn Syst Meas Control 121:566–571
43. Piezoelectric Ceramics Data Book for Designers, Morgans Electroceramics
44. Mateu L, Moll F (2005) Optimum piezoelectric bending beam structures for energy harvesting using shoe inserts. J Intell Mater Syst Struct 16(10):835–845
45. Siores E et al (2011) An investigation of energy harvesting from renewable sources with PVDF and PZT. Smart Mater Struct 20(5):055019
46. Sodano A et al (2005) Generation and storage of electricity from power harvesting devices. J Intell Mater Syst Struct 16
47. Mohammadi F, Khan A (2003) Power generation from piezoelectric lead zirconate titanate fiber composites. Mater Res 736:1–6
48. Funasaka T et al (1999) Piezoelectric generator using a LiNbO3 plate with an inverted domain. In: IEEE Ultrasonics Symp, Honolulu, HI, pp 959–962
49. Griffith BP et al (1995) A gait powered autologous battery charging system for artificial organs. Am Soc for Artif Internal Organs J 41M:588–595
50. Pondrom P et al (2014) Vibration-based energy harvesting with stacked piezoelectrets. Appl Phys Lett 104(17):172901
51. Gershenfeld N et al (1998) Parasitic power harvesting in shoes. In: Proceedings - 2nd IEEE International Conference on Wearable Computing (California), pp 132–139
52. Granstrom J et al (1810) (2007) energy harvesting from a backpack instrumented with piezoelectric shoulder straps. Smart Mater Struct 16(5)
53. Kymissis J et al (1998) Parasitic power harvesting in shoes. In: Proceedings of the second international symposium on wearable computers. Digest of Papers, IEEE, pp 132–139
54. Corina Covaci and Aurel Gontean (2020) Piezoelectric energy harvesting solutions: A review. Sensors 20(12):3512
55. Roundy S et al (2003) A study of low level vibrations as a power source for wireless sensor nodes. Comput Commun 26:1131–1144
56. Kim et al (2013) Comparison of MEMS PZT cantilevers based on d_{31} and d_{33} modes for vibration energy harvesting. J Microelectromech Syst 22(1)
57. Renaud M et al (2009) Harvesting energy from the motion of human limbs: the design and analysis of an impact-based piezoelectric generator. Smart Mater Struct 18(3):035001

58. Zhang Y, Cai C (2012) A retrofitted energy harvester for low frequency vibrations. Smart Mater Struct 21(7):075007
59. Pillatsch et al (2012) A scalable piezoelectric impulse-excited energy harvester for human body excitation. Smart Mater Struct 21(11):115018
60. Sohn J et al (2005) An investigation on piezoelectric energy harvesting for MEMS power sources. Proc. IMechE 219 part C. J. Mech. Eng. Sci.:429–436
61. Todaro et al. (2017) Piezoelectric MEMS vibrational energy harvesters: advances and outlook. Microelectron Eng, 183: 23–36
62. Ammar et al (2005) Wireless sensor network node with asynchronous architecture and vibration harvesting micro power generator. In: Proceedings of the 2005 joint conference on Smart objects and ambient intelligence: innovative context-aware services: usages and technologies. ACM, pp 287–292
63. Wright et al (2003) A study of low level vibrations as a power source for wireless sensor nodes. Comput Commun 26:1131–1144
64. Shen et al (2009) Micromachined PZT cantilever based on SOI structure for low frequency vibration energy harvesting. Sensors Actuators A Phys 154(1):103–108
65. Liu et al (2008) A mems-based piezoelectric power generator array for vibration energy harvesting. Microelectron J 39(5):802–806
66. Marzencki et al (2005) Design and fabrication of piezoelectric micro power generators for autonomous microsystems. In: Proc. Symp. On design, test, integration and packaging of MEMS/MOEMS DTIP05, Montreux, Switzerland, pp 299–302
67. Kim et al (2005) MEMS power generator with transverse mode thin film PZT. Sensors Actuators A 122:16–22
68. Yang et al (2009) Power generation with laterally packaged piezoelectric fine wires. Nat Nanotechnology 4(1):34–39
69. Xu et al (2010) Piezoelectric-nanowire-enabled power source for driving wireless microelectronics. Nat Commun 1:93
70. Dagdeviren et al (2014) Conformal piezoelectric energy harvesting and storage from motions of the heart, lung, and diaphragm. Proc National Academy Sci 111(5):1927–1932
71. Dogheche et al (2005) A bi-stable micro-machined piezoelectric transducer for mechanical to electrical energy transformation. In: Proceeding Symposium on Design, Test, Integration and Packaging of MEMS/MOEMS, Montreux, Switzerland, pp 303–304
72. Kim et al (2005) Piezoelectric energy harvesting under high pre-stressed cyclic vibrations. J Electroceram 15(1):27–34
73. Adhikari et al (2009) Piezoelectric energy harvesting from broadband random vibrations. Smart Mater Struct 18, No. 11
74. Lang et al (2001) Vibration-to-electric energy conversion. IEEE Trans Very Large Scale Integration (VLSI) Systems 9(1):64–76
75. Meninger S (1999) A low power controller for a MEMS based energy converter MSc thesis. Massachusetts Institute of Technology
76. Elvin and Erturk (2013) Advances in energy harvesting methods. Springer Science & Business Media
77. Sue C-Y, Tsai N-C (2012) Human powered MEMS-based energy harvest devices. Appl Energy 93:390–403
78. Roundy S (2003) Energy scavenging for wireless sensor nodes with a focus on vibration to electricity conversion. PhD Thesis, University of California, Berkeley
79. Liu et al (2011) Design of Self-powered Digital Over-current Protector. IEEE, pp 1047–1050
80. Moll F et al (2020) Mechanical energy harvesting taxonomy for industrial environments: application to the railway industry. IEEE Trans Intell Transp Syst 21(7):2696–2706
81. Ali et al (2019) Piezoelectric energy harvesters for biomedical applications. Nano Energy 57: 879–902
82. Surmenev et al (2019) Hybrid lead-free polymer-based scaffolds with improved piezoelectric response for biomedical energy harvesting applications: A review. Nano Energy 62:475–506
83. Mhetre et al (2011) Micro energy harvesting for biomedical applications: a review. IEEE 3rd Internat Conf Electron Comput Technol 3:1–5

Energy Harvesting Techniques and Trends in Electronic Applications

Pavan Mehta, Anupama Gaur, Chandan Kumar, Anveshkumar Nella, Anirban Bhowmick, and Maheswar Rajagopal

1 Introduction

Recent development in electronics and wireless technology has open doors to the advancement of low-power electronics [1] and wireless sensors [2] to find its usefulness in a different application. The wireless nature of some systems requires independent or their own power source for its working which are usually powered by a battery with a limited power supply. Power when needed in large increases equipment batteries size as well as the cost of the devices which in turn increases the burden of replacing or charging them [3]. Therefore, more and more efforts are being made to shift nonrenewable to new long-lasting and renewable energy sources to satisfy the energy requirements of these systems. Other ways include (a) using a nearby existing higher energy source, (b) using an active source for wireless power

P. Mehta (✉)
Research Scholar, School of EEE, VIT Bhopal University, Sehore, India
e-mail: pavan.mehta2020@vitbhopal.ac.in

A. Gaur
Research Scholar, School of Material Science and Technology, IIT (BHU), Varanasi, India
e-mail: anupama.rs.mst14@itbhu.ac.in

C. Kumar
School of Engineering and Technology, Central University of Rajasthan, Ajmer, India
e-mail: ckumar.rs.bme13@iitbhu.ac.in

A. Nella · A. Bhowmick
School of EEE, VIT Bhopal University, Kothri Kalan, Sehore, India
e-mail: nellaanvesh@gmail.com; anirban.bhowmick@vitbhopal.ac.in

M. Rajagopal
Department of ECE, Centre for IoT and AI (CITI), KPR Institute of Engineering and Technology, Coimbatore, India
e-mail: maheshh3@rediffmail.com

A. Nella et al. (eds.), *Energy Harvesting Trends for Low Power Compact Electronic Devices*, EAI/Springer Innovations in Communication and Computing,
https://doi.org/10.1007/978-3-031-35965-1_11

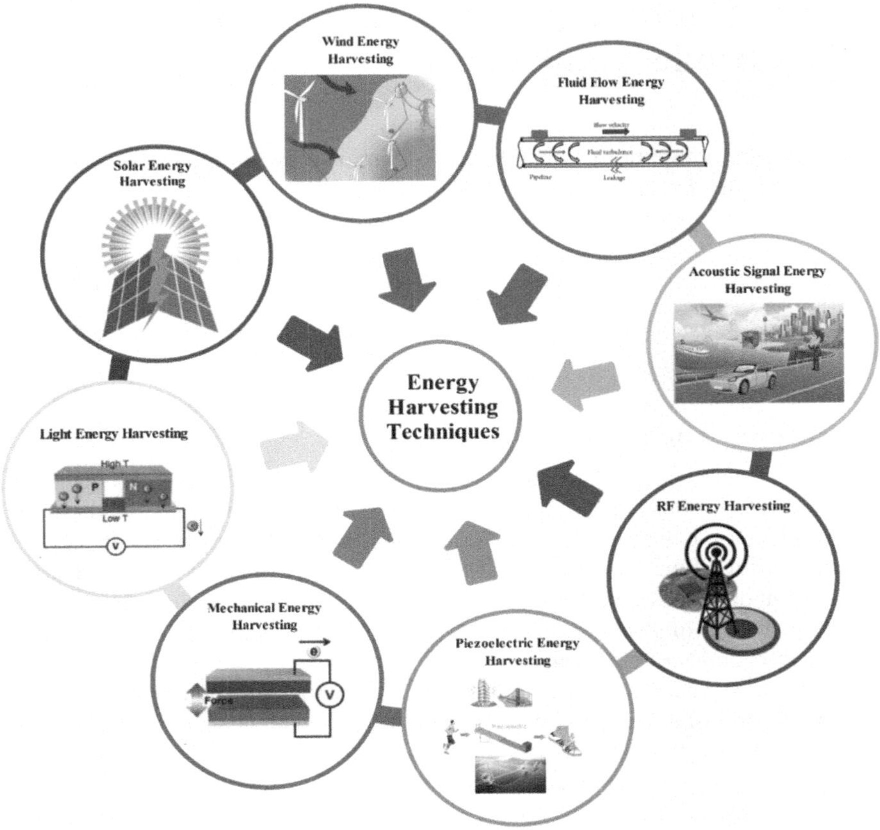

Fig. 1 Different sources of energy harvesting

generation, or (c) efficiently harvesting energy from the environment. Thus, the basic idea behind energy harvesting (EH) is to recover unused or excess energy, which is the by-product of human activities, natural or some automated processes generally unused and hence considered to be easily available abundant energy in the environment. Thus, energy harvesting or sometimes called energy scavenging is a concept to recover ambient energy from sources in the environment. Utilizing this easily available ambient energy sensor, wireless nodes and portable electronic devices can be made independent, eventually eliminating the need for battery and its maintenance. Examples of ambient sources include electromagnetic energy from solar, infrared, or radio frequencies; energies in the form of mechanical energy due to vibration, tension, and stress; heat energy from internal combustion engines, furnaces, and other heat sources; other energy from human activities like various metabolic processes and functions inside or outside the human body [4–7]; etc., as shown in Fig. 1. Sensors depending upon various energy sources are used to detect energy from abovementioned sources and then transformed into usable form [8], i.e., electrical power, voltages, and currents. The transformed signals can be harvested, stored, and further managed according to the power necessities of portable electronics and wireless sensor networks (WSNs) [9–12].

When we talk about energy harvesting, it should be known that the primary objective of EH is not only to generate energy in the bulk, but to store the obtained energy so that we can use it later. In short, we can say that the key purposes of energy harvesting technology may be but not limited (1) to eliminate the main power lines, (2) to reduce or eliminate the use of batteries, (3) to increase the lifetime, (4) to increase and maintain functionality, (5) easy installation, (6) lower cost, and (7) waste prevention [13].

2 Types of Energy Harvesting and Applications

The classification of numerous energies harvesting system is discussed in this section. The energy harvesting systems are broadly divided into macro and micro energy harvesting systems depending upon the amount of power harvested. Some common harvesting systems are wind energy harvesting, solar energy harvesting, acoustic signal energy harvesting, radio frequency energy harvesting, piezoelectric energy harvesting, mechanical energy harvesting, light energy harvesting, and thermal energy harvesting. In this section, wind energy harvesting is reported first, followed by other techniques.

2.1 Wind Energy Harvesting

Humans through ages use wind energy for various purposes to their benefits; it is a conventional way of energy collection. Wind speed and direction change due to altitude, terrain, and vegetation and on water bodies. This kinetic energy generated because of wind flow, which is collected by large wind turbines, can be harvested for the production of electricity on large scale [1]. However, small wind turbines can be used on highways and bridges by the sea, where the wind pressure is quite high and from which energy can be produced together with the speed generated by the movement of vehicles. The most common application of this can be to give power to traffic lights and signals and also to power highway lights and signs. Another alternative is the wind power in combination with other sources can supply energy for low-power applications.

Park et al. explored the practicability of micro wind generators on a cable-stayed bridge in order to provide power to the wireless sensors attached on it. The turbine power is determined by wind tunnel test by using three different rotors. The turbine with six blades having diameter of 13.8 cm operating at wind speed of 7 m/s is able to generate maximum power of 38.79 mW with the power density of 2.93 mW/cm^2

[14]. Perez et al. presented a wind harvester that combines flutter effects with electret-based electrostatic conversion technology. The model presented for this study consists of two parallel copper plates having length and thickness of 50 mm and 200μm, respectively. Copper foils are coated with a 25-μm-thick Teflon polytetrafluoroethylene (PTFE) produced by corona discharge. In addition, a 25-μm-thick polyvinylidene fluoride (PVDF) film with a 10-nm-thick layer of gold deposited on both sides is attached to the bluff body. As the wind blows the membrane up and down, the power of the energy collector changes continuously. The membrane starts shaking as the wind speed approaches 10 m/s.

Experimental data shows that the designed wind harvester was able to produce 200 V output voltage and 2.1 mW power when the electret voltage is −650 V with a wind speed of 30 m/s [15]. Chen et al. proposed a triboelectric flutter energy harvester model which can operate on wind speed at 1.6 m/s to generate 175 V of output voltage, current of 434μA, and power of 8 m/s with a power density up to 52 $nWcm^{-3}$ [16].

2.2 Harvesting Solar Energy

Solar energy is also abundant and free source present everywhere around us. Photovoltaic cells or sheets can be installed where the intensity of incoming solar energy is sufficient to capture it. Earth atmosphere receives solar energy on average approximately 1018 kWh/year and the power density approximately 1353 kW/m^2 on its top. The energy collected varies from 100 mW/cm^2 direct sun to 100 W/cm^2 [17]. Gajanur et al. present techniques to maximize the power of solar cells by means of maximum power point tracking (MPPT) method. Here, the input environments are continuously monitored, and output power is determined on the equivalent load from the I-V curve of the solar cell. However, sun is abundant, but due to different reasons, its availability and intensity are limited, so supplementary storage devices may or may not be required to produce continuously harvested energy. Gajanur et al. presented a tracking system that detects obstacles up to 2 km on the railway track. The proposed model consists of a CMOS (complementary metal oxide semiconductor) camera mounted forward-facing the train with a Wi-Fi module, a microcontroller, and a liquid crystal display (LCD) screen fitted to all the carriage for the passenger to detect the track situations so to avoid any unexpected condition by pressing the emergency brakes. Lithium-ion batteries which were powered by solar panels are mounted on the carriage roofs. Improved triple junction (ITJ) solar cells with a nominal efficiency of 26.8% regulate the voltage between 0.3 V and 5.5 V at regular intervals to maximize performance. The DC-DC converter is designed to distribute the current to different rails, 3.3 V and 2.8 V to the bus, to power the Wi-Fi module, the microcontroller, and the camera, respectively [18]. Silva-Leon et al. presented a novel design that uses wind and solar energy. Piezoelectric and flexible photovoltaic material strips were used to track the idea of inverter flag. Total power generated by the system accounts 3–4 mW at wind speed ranging from 0 to 26 m/s and 1.8kLux constant light irradiation [19]. Toh et al. design a circuit for flexible power generation. The power circuit generated 56 μW of electrical energy, and the system was capable of monitoring the wearer's health status, such as body temperature, and then transmitting it to the center wirelessly. Estimated operating time exceeds 15 hrs for the whole system [17].

2.3 Energy Harvesting from Fluid Currents

Air and water currents, whether natural or artificial, are available everywhere from rivers to wind and to water pipes and ventilation shafts. Although these fluids and their flows seem to live in completely different environments, very similar methods can be used to extract energy from fluid flows using rotors and turbines to convert mechanical energy into electrical energy through electromagnetic induction or piezoelectric methods [20]. Wang et al. presented a model based on the principle of electromagnetic induction using oscillations caused by fluid flow in a close pack environment. The prototype consists of flow channel with two copper tubes, a polyester film (PE) attached to the closed channel, and a permanent magnet attached to the PE film. The permanent magnet was walled through a conductive coil that is passed round the inside casing of the coil, which was held by the outside of the shell. A polyester membrane attached to an integrated permanent magnet enclosed by a fixed coil is oscillated by the fluid gradient, i.e., difference in pressure. The vibrational energy is then transformed into electrical energy by coil induction. This configuration produced an output voltage up to 11 mV for oscillations at 30 Hz [20].

D. Wang et al. developed an energy harvesting apparatus for oscillating fluid lines having tube diameter up to 2 mm. The effort is on development of a mathematical model-based vibrational energy harvesting mechanism. He showed experimentally that the developed model produces electricity by collecting current-induced oscillations by using piezoelectric membrane and electricity was generated as the result of the pressure fluctuations by the flow of fluid in the tube. The size of the piezo-membrane is 13×25 mm^2. The opening voltage of the energy harvester is 2.2 V, and the instantaneous output power is 0.2μW with an excitation pressure of 1.196 kPa and a frequency of 26 Hz. The model best suits for the households or a small subsystem because of its low energy requirements [21]. Azevedo et al. investigated usefulness of various small water and wind turbines with different flow characteristics to power wireless sensor nodes. Turbines were varied according to the working location, geometric assembly, blade size, length, and its number along with other characteristics. Depending upon the designing and flow rate of turbine, the power density of the wind turbines varied between 0.01 and 5.88 mW cm^{-3}. The power density of the water turbines varied from 0.63 to 70.1 mW cm^{-3} conditional to design and flow rate of the turbine. Flow rate plays an important role in the turbine energy conversion efficiency; also the increase in the flow rate may not increase the efficiency [22].

2.4 Energy Harvesting from Acoustic Signals

Energy harvesting models or hypothesis mainly focused on mechanical vibration, light, and temperature gradients energy sources. Exploring acoustic energy as a viable source is promising due to its abundance in our surroundings. Acoustic noise, generated by vibrating sources, produces pressure waves that the human ears can detect within the frequency range of 20 to 20,000 Hz. These detected waves are then converted into corresponding electrical signals [2]. Sound energy is present in the environment almost continuously and to a significant extent, e.g., on railways, runways, shipyards, or roads (vehicle engine noise and sirens), loud music noises on parties and discos, construction and building sites, and likewise sources. Other sources such as cell phone ring tone and dial tune (80 dB), jet engines (10 dB), train horns (90 dB), arm exercises (98 dB), and airplanes (19 dB) [23] generate enough sound pressure levels (SPL) for power generation. Runway lighting using aerodynamic noise energy would be an innovative idea.

Iizumi et al. presented model of an acoustic energy harvester (AEH) with a piezoelectric thin-film membrane on a 300-μm-thick silicon wafer as substrate. This harvester has two upper electrodes and one lower electrode of piezoelectric membrane. Silicon wafer is oxidized to SiO_2 (1.5μm); a Ti layer (100 Nm) and then a Pt layer (100 Nm) are deposited and patterned to make the bottom electrode as shown in Fig. 2. Above Pt/Ti electrode, a PZT layer of about 1μm is deposited, and the top electrode is made by depositing an Al layer of about 100-nm thickness. Finally, the back of the silicon wafer is etched with an increased intracranial pressure (ICP) to reveal the film. During production upper electrodes are electrically connected in series. Maximum power of approx. 82.8 pW was delivered by the harvester with internal impedance of 50 Ω and a resonant frequency of 0.92 kHz at 100 dB SPL (sound pressure level) [24]. Yang et al. proposed a triboelectric acoustic energy harvester model by using a Helmholtz resonator. The separated multilayer consists of a thin layer of nanoporous aluminum and a thin coat of polytetrafluoroethylene (PTFE) on which copper is deposited. Aluminum (al) thin film acts as the connection surface, the electrode, and the PTFE layer acts as the second electrode. As the sound wave hits, the resonator magnifies the incoming sound force, and the PTFE membrane vibrates, repeating the sequence of contact and separation. The maximum electrical power density of 60.2 mWm^{-2} generates to

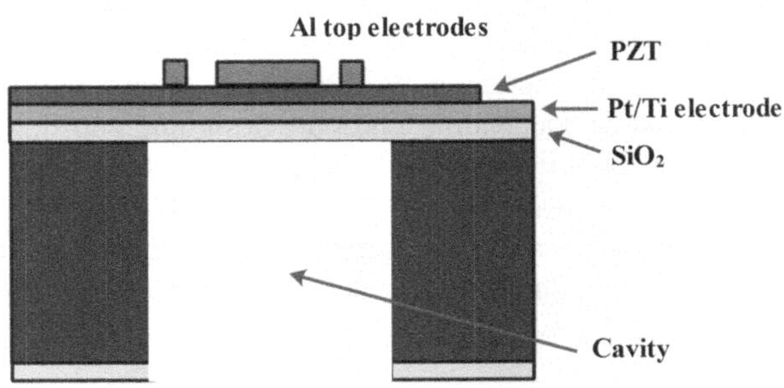

Fig. 2 Schematic cross-sectional PZT energy harvester [24]

approximately 376μW of power at 110 dB SPL [25]. Khan et al. proposed that the conical cavity Helmholtz resonator has a high acoustic stiffness when compared with the conventional Helmholtz resonator, the resonance frequency of the increase in pressure with the increase of the conical cavity. This developed acoustic energy harvesting device produced a maximum power of 214.23μW at 130 dB SPL and a bandwidth of 1453–1542, 1710–1780, and 1848–1915 Hz [23].

2.5 Radio Frequency Energy Harvesting

Radio frequency (RF) is everywhere from satellite and radar or in radio and TV stations, Wi-Fi routers, and other communication networks. However, energy or frequency levels may differ due to reasons such as landscape, number of device handlers, network congestion, etc. In such environmental situations, big broadband antennas with large gain, scalability, and easy installation are required to generate useful electrical energy. In addition, it is tough to distinguish between an efficient RF signal and an inefficient RF signal. Based on abovementioned factors, there is little actual field collection of RF energy. However, the RF output power is very low, on the order of 1μW/cm^2 [26]. However, with methods involving wideband frequency bands and automatic frequency tuning, output power may be improved; thereby RF can be a potential energy harvesting source. A detailed block diagram of the radio frequency energy harvesting system is shown in Fig. 3. Koohestani et al. presented a printed rectifier antenna tuned frequency at 2.45 GHz for RF energy harvesting; its direct dimensions are 24.9 × 8.6 × 1.6 mm^3. At load resistance 4.7 kΩ, the output power is 97 mV and at power − 20 dB the maximum RF conversion efficiency is 20%. A rectifier antenna successfully powered a 3 V LED [27]. Shen et al. proposed an L-shaped two-port three-band rectifier antenna for RF energy harvesting. Rectenna operating frequencies are GSM-900, GSM-1800, and UMTS-2100. The efficiency of the RF harvester was above 40%, the output voltage was 600 mV, and the power density was 500μW/cm^2 [26]. Alex-Amor et al. investigated a capacitor-combined RF collector device for power enhancement. The system configuration includes an RF Archimedean spiral antenna with a Cockcroft-Walton multiplier

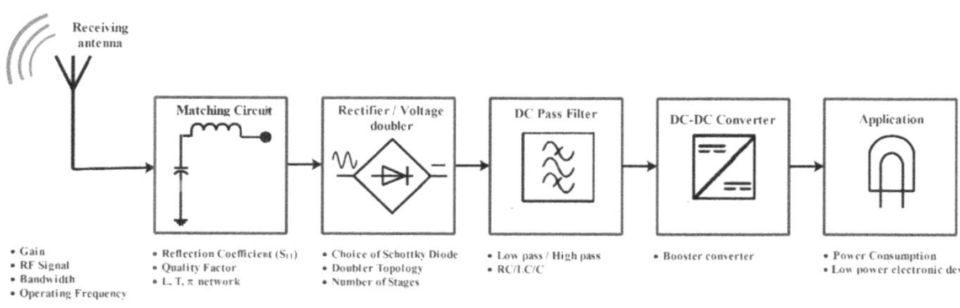

Fig. 3 Block diagram of a radio frequency energy harvesting system

circuit. The antenna has demonstrated that it can operate on frequencies ranging 350 MHz to 16 GHz with exceptional imaging. The frequency offset of the layout circuit design changed to account for the different parasitic components of the PCB and units. At an input power of 0 dBm, the modified efficiency of the circuit was 30%. The charge on the capacitor was more than 1.25 V, which is sufficient for the operation of the temperature sensor [28].

2.6 *Piezoelectric Energy Generation*

When mechanically loaded, materials of special class, i.e., piezoelectric materials, generate an electric potential gradient that results in an electric field proportional to the applied voltage. The work [29] depicts behavior of the piezoelectric material on the application of load, compared to modeling with time. By converting motion or vibration into an AC wave, this action can be used as a transduction process in vibration energy harvesters. In a conventional piezoelectric power generator, the vibrating part of the device is a cantilever beam made of piezoelectric material to which a proof mass is put on it to reduce the resonant frequency. Booms with low stiffness are needed to absorb low-frequency vibrations. Similarly, bimorph structures, in which piezoelectric materials are put over one another as two layers, are often cast-off to increase the effectiveness of harvesters because they allow greater loading at a resonant frequency lower than single-layer structures [21].

Yoon et al. used Volture V22BL, a commercial piezoelectric harvester, which generated 3.072 mW output power at 1 g acceleration and vibration source of 53 Hz. However, it failed because the unsteady operating environment mostly failed to match with the resonant frequency of the system, drawing the attention to tune the resonant frequency [30]. Yang et al. showed a battery-less piezoelectric energy harvesting device that was able to deliver a total energy of 160μJ at 2.5 V to transmit a signal to a transceiver over a short distance by pressing a single button [31]. Lee et al. presented a piezoelectric energy harvester to scavenge energy from the speeding up of tires while accelerating as shown in Fig. 4. This system produced 13.7 nW cm^{-3} over a 60-mm field, which was enough to continuously power the node in spite of its low-power density [29].

2.7 *Mechanical Energy Harvesting*

Environment is in abundance of mechanical sources which offers a world of opportunities for energy harvesting. Attempts made on mechanical scavenging were able to generate power ranging from 4 W to 800 W [6]. Mechanical energy sources can come in different forms such as balanced source and periodic and vibration source, and the best results come from vibration source. Wang et al. introduced an electromagnetic vibration collector for industrial machine vibration,

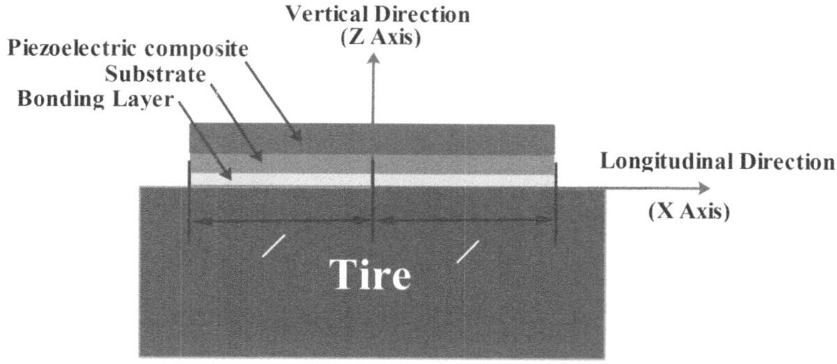

Fig. 4 Concept of a piezoelectric energy harvester [29]

i.e., an air compressor, and which was able to generate 1.56 mW of machine vibration with accelerations of 25–48 mg and frequencies of 49.3–49.7 Hz [32]. Sari et al. projected an electromagnetic generator proficient of scavenging low-frequency ambient oscillations by translating them to high-frequency oscillations. The proposed generator efficiently harvested energy from 70–150 Hz ambient oscillations and provided a voltage of 0.57 mV and power of 0.25 nW with a generator size of 0.15 $cm^{3.}$ Power and voltage can be increased by increasing the number of lings or coil turns [33]. Gaur et al. developed a polymer-based piezoelectric device made up of bio-waste polymer composite. The device can produce voltage of around 65 V and a power density of 84μW/$cm^{2.}$ The device was also able to generate power through various mechanical movements of human motion such as walking, wrist twisting, finger tapping, and knee bending [34].

2.8 Energy Harvesting from Light

Light is an electromagnetic radiation in a specific part of the electromagnetic spectrum. Light which is visible is generally well-defined as light with a wavelength between 400 and 700 nanometers (nm), which is between infrared (longer wavelengths) and ultraviolet (shorter wavelengths). That wavelength implies a frequency range of approximately 430–750 terahertz (THz) [35–37].

Yu et al. proposed a new power harvesting and power management circuit that includes a maximum power point tracking (MPPT) circuit, an energy storage circuit, an instantaneous energy discharge circuit, and a DC-DC boost converter. It harvests energy from the dark interior, causing the harvester to run near the maximum power point and providing power to the sensor node. Measurement result shows that the prototype circuit can scavenge energy from interior light sources with an input power of 72.74μW and efficiently supply power to a smart wireless sensor node of temperature and humidity with a power consumption of 105 mW per duty cycle [35]. Iyer et al. proposed a laser-based system for in-room charging of smartphones

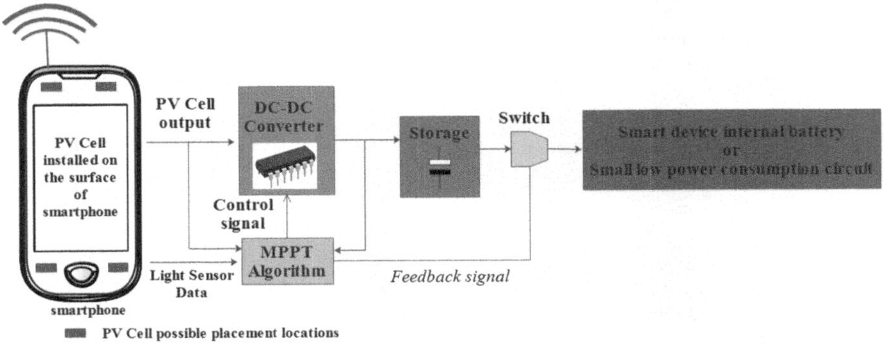

Fig. 5 Proposed light energy harvesting system using PV cells [37]

and IoT devices. This system is the first complete wireless power solution capable of delivering power safely across a room and offers extensive characterization backed by empirical data that shows our system is never exposed to harmful high-power lasers. The results of this system show that it can transmit more than 2 W of power at a distance of 4.3 meters and 12.2 meters with a smartphone (25 cm^2) and a desktop receiver (100 cm^2). Furthermore, extensive characterization of our safety system shows that we can turn off our laser source long before a person moving at a maximum of 4 m/s even comes within range of a high-powered laser beam [36]. Jabbar et al. developed an energy management system that includes an ambient light energy harvesting method and introduce various systems that use solar cells to share and complement the energy consumption of the battery as shown in Fig. 5. Ambient light energy harvesting and its intensity can lead to light energy harvesting technology in low-power devices. A new system for using solar concentrators in smartphones will be launched to improve the use of light energy in smartphones. It has been shown that solar energy can distribute battery consumption in smartphones, wearables, and IoT devices using a small and low-power logic and algorithm in a smart wearable device. The PV cells used in this study were very small ($22 \times 7 \times 1.8$ mm^3) and had a breakdown voltage of 1.89 V and a short-circuit current of 15 mA and a typical efficiency of 22% with standard experimental values of $\lambda = 1000$ W/m^2 and a temperature of 25 °C [37].

2.9 Extraction of Thermal Energy

Thermal or heat energy is the heat generated in the environment or the heat produced through certain process or activities going in the nature. Either thermoelectric or pyroelectric effects are applied to recover these forms of energy. Thermoelectric effects such as the Peltier effect, the retrograde effect, and the Thomson effect can produce power as extensively as these heat sources are present. The thermal gradient, i.e., the heat or temperature difference between the source and environment, plays a vital role in the conversion efficiency; the larger the temperature difference, the better the performance [38]. The pyroelectric effect, which is created on the spontaneous polarization of certain anisotropic solids due to a time-dependent temperature change, can be used if the medium has a uniform temperature gradient, i.e., no large temperature fluctuations. The power output of thermal fluctuations is approximately 10μW/cm^2 and can be up to 60 W/cm^2 with the thermoelectric source. Although this technique is scalable and robust, the conversion efficiency and output voltage are low because the temperature gradients are typically small [39]. Gao et al. proposed an idea to utilize the normal temperature difference between the bottom of the rail and the ground or the ballast under the rail on Chinese lines. Experimentally he showed that the temperature difference of up to 30 °C is able to generate electricity with a thermoelectric generator (TEG). The electricity produced by the prototype can be amplified with a DC-DC amplifier and then stored in lithium (li) batteries. The harvester produced 5.8 mW of power with a temperature difference of 8 °C and an impedance of 7 Ω in the field tests. In laboratory experiments, a power of 316.8 mW was obtained at a temperature gradient of 29.2 °C and a resistance of 9.6 Ω at a voltage of 1.76 V [40].

Ahn et al. have discussed a thermal energy harvester for use in high-speed vehicles. The harvester could generate sufficient energy to uninterruptedly power either 2 three-axis accelerometers or 20 to 30 temperature sensors. This device uses the temperature variance among the wheelset bearing housing and the environment using a thermoelectric module (TEM), i.e., the same as a thermoelectric generator (TEG). Laboratory tests presented that the device can provide a maximum power of 27.97 mW at 0.246 V for a temperature difference of 30 °C and at a load resistance of 2.197 Ω [38]. Becker et al. proposed an autonomous sensor system for keeping track of strain developed in the advanced aircraft materials. Thermoelectric generators (TEGs), solar cells, and vibration generators are being recognized as energy harvesting methods. In the laboratory test conducted, it was found that at a thermal gradient of 70° C, the power produced by the TEG was about 20 mW, which is sufficient for independent operation of the sensor node and associated data communication protocols [39].

3 Comparative Study

A detailed comparison of different types of energy harvesting techniques, advantages, disadvantages, and applications is discussed below (Table 1).

For electronic devices, the energy harvesting technique must be feasible in versatile form, portable, easy to operate, and energy-dense. This table discusses each energy source and the technology used to harvest it, as well as the benefits and drawbacks of these technologies, the range of power density obtained, and the application domains that can profit from using them to construct self-powered electronic devices. Depending on the nature of the energy source and the viability

Table 1 Comparison of various energy harvesting techniques [6]

Energy source	Characteristic	Harvested power	Advantages	Limitations	Application
Wind energy harvesting	Environment	~100μW/cm^3	Remote wireless sensors Maturity of wind technology in other fields Wind is constantly blowing	Difficult installation at every place Low-efficiency wind conditions are not easily predictable	
Solar energy harvesting	Environment	10–100 mW/cm^2 (outdoor) <10μW/cm^2 (indoor)	High-output voltage Low fabrication costs Predictable	Unavailable in light Not controllable	Environment monitoring
Fluid flow energy harvesting			Mature technology	Difficult to apply form factor	
Acoustic signal energy harvesting	Machines	~10μW/cm^3	Small handheld electronic devices		
RF energy harvesting	Antenna	0.001μW/cm^2 - 0.1μW/cm^2	Available anywhere, anytime Predictable Controllable	Distance dependent Low-power density	Environment monitoring
Piezoelectric energy harvesting		4–330μW/cm^3	High-power density No external voltage source Simplicity design and fabrication Controllable	Highly variable output Unpredictable	Infrastructure Monitoring Automotive Healthcare
Mechanical energy harvesting	Human	~4μW/cm^3	Handheld electronic devices or remote wireless actuators		
	Machines	~800μW/cm^3			
Light energy harvesting	Outdoor	100 mW/cm^2			

(continued)

Table 1 (continued)

Energy source	Characteristic	Harvested power	Advantages	Limitations	Application
	Indoor	100μW/cm^2	Handheld electronic devices		
Thermal energy harvesting	Human	60μW/cm^2	Remote wireless sensors and actuators	Low cost Good efficiency Easy installation on axlebox and bogie Good power output	Power output depends on time of the year and weather Optimal load resistor depends on thermal gradient Cooling system may be needed
	Industrial	~1–10μW/cm^2			

of the associated conversion method, each can be adapted to specific functions in sustainable energy systems.

For example, wind energy harvesting system, a mature technology for all fields, utilizes consistently flowing wind and can generate power density in the range of μW/cm^3, but dependent on the geographical, climate, and environmental conditions. Energy harvesting from solar power produces a power density of μW/cm^2 under indoor lighting conditions and mW/cm^2 under outdoor lighting conditions. The amount of harvested energy depends on the weather and requires constant illumination. Acoustic energy harvesting performs better than the others. In contrast to the power produced by piezoelectric generators, triboelectric-based thin-film nanogenerators have demonstrated a good power density with an acoustic pressure range of 70–110 dB.

The fundamental benefit of RF harvesters is the accessibility of their power source and does not require any sort of transducer material. It has low-power densities in the range of nW/cm^2. The sizes, shapes, and types of diodes used in the rectifier block and impedance matching network present design issues for RF harvesters. Piezoelectric harvesters are widely employed in a variety of low-power applications. They are small in size, have a high output voltage, and just need simple excitation. Piezoelectric technology faces difficulties with downsizing, frequency dependence of the harvester, and operating in the human frequency range. The power density of TEG is in the range of mW/cm^2, and it uses abundant heat energy to generate power. TEGs have the potential to power electronic and electrical equipment like low-energy LED lights by harvesting low-grade heat energy from factory manufacturing processes and exhaust heat from internal combustion engines.

The mentioned harvesters have been successfully used in a number of wearable and implantable low-power applications, but there are still a number of obstacles to be overcome before we can completely rely on energy harvesting without the use of batteries. The harvesters' resilience, biocompatibility, miniaturization, flexibility, improvement of the material characteristics, development of effective and low-power management circuits, and enhancing the harvesters' energy conversion efficiency are the challenges.

4 Conclusion

This chapter discussed a variety of energy generation technologies, including wind, solar, fluid, sonic, high frequency, piezoelectric, mechanical, light, and thermal energy. Their advantages, limitations, and applications were also discussed. The use of energy harvesting systems will become even more important as they are increasingly adapted to electronics, as their applications have environmental, logistical, and economic advantages. These systems can extend the useful life and independence of equipment. However, the energy harvesting system of each application depends on the application itself and the environment in which it is used. In addition, the performance of an energy harvesting system depends on many physical factors and the combination of collection and storage methods. Therefore, it is worth considering that within the existing methods and physical conditions, it is possible to recover the greatest amount of usable energy.

References

1. Hesham R, Soltan A, Madian A (2021) Energy harvesting schemes for wearable devices. AEU Int J Electron Commun 138:153888
2. Kim KK, Choi J, Ko SH (2021) Energy harvesting untethered soft electronic devices. Adv Healthc Mater 10(17):2002286
3. Bosso N, Magelli M, Zampieri N (2021) Application of low-power energy harvesting solutions in the railway field: a review. Veh Syst Dyn 59(6):841–871
4. Grossi M (2021) Energy harvesting strategies for wireless sensor networks and mobile devices: a review. Electronics 10(6):661
5. Sanislav T, Mois GD, Zeadally S, Folea SC (2021) Energy harvesting techniques for internet of things (IoT). IEEE Access 9:39530 –39549
6. Williams AJ, Torquato MF, Cameron IM, Fahmy AA, Sienz J (2021) Survey of energy harvesting technologies for wireless sensor networks. IEEE Access 9:77493 –77510
7. Choi J, Jung I, Kang C-Y (2019) A brief review of sound energy harvesting. Nano Energy 56: 169 –183
8. Mateu L, Moll F (2005) Review of energy harvesting techniques and applications for microelectronics. VLSI Circuits and Systems II 5837:359 –373
9. Nabavi S, Zhang L (2016) Portable wind energy harvesters for low-power applications: a survey. Sensors 16(7):1101
10. Calautit K, Nasir DS, Hughes BR (2021) Low power energy harvesting systems: state of the art and future challenges. Renew Sust Energ Rev 147:111230

11. Davidson J, Mo C (2014) Recent advances in energy harvesting technologies for structural health monitoring applications. Smart Mat Res
12. Ferdous MR, Reza AW, Siddiqui MF (2016) Renewable energy harvesting for wireless sensors using passive RFID tag technology: a review. Renew Sust Energ Rev 58:1114–1128
13. Pozo B, Garate JI, Araujo JA, Ferreiro S (2019) Energy harvesting technologies and equivalent electronic structural models. Electronics 8(5):486
14. Park J-W, Jung H-J, Jo H, Spencer BF Jr (2012) Feasibility study of micro-wind turbines for powering wireless sensors on a cable-stayed bridge. Energies 5(9):3450–3464
15. Perez M, Boisseau S, Gasnier P, Willemin J, Reboud J-L (2015) An electret-based aeroelastic flutter energy harvester. Smart Mater Struct 24(3):035004
16. Chen X, Ma X, Ren W, Gao L, Lu S, Tong D, Wang F, Chen Y, Huang Y, He H, Tang B, Zhang J, Zhang X, Mu X, Yang Y (2020) A triboelectric nanogenerator exploiting the Bernoulli effect for scavenging wind energy. Cell Rep Phys Sci 1(9)
17. Toh WY, Tan YK, Koh WS, Siek L (2014) Autonomous wearable sensor nodes with flexible energy harvesting. IEEE Sensors J 14(7):2299–2306
18. Gajanur NR, Singh A, Jain A (2016) Solar powered railway track monitoring system. In: IEEE International Conference on Power and Renewable Energy (ICPRE), pp 190–194
19. Silva-Leon J, Cioncolini A, Nabawy MR, Revell A, Kennaugh A (2019) Simultaneous wind and solar energy harvesting with inverted flags. Applied Energy 1(239):846–858
20. Wang DA, Chang KH (2010) Electromagnetic energy harvesting from flow induced vibration. Microelectron J 41(6):356–364
21. Wang DA, Ko HH (2010) Piezoelectric energy harvesting from flow-induced vibration. J Micromech Microeng 20(2):025019
22. Azevedo JAR, Santos FES (2012) Energy harvesting from wind and water for autonomous wireless sensor nodes. IET Circuits, Devices Syst 6(6):413–420
23. Khan FU (2016) An improved Design of Helmholtz Resonator for acoustic energy harvesting devices. International Conference on Intelligent Systems Engineering:2–7
24. Iizumi S, Shu K, Tomioka S, Tsujimoto K, Uchida Y, Tomii K, Matsuda T, Nishioka Y (2011) Lead zirconate titanate acoustic energy harvesters utilizing different polarizations on diaphragm. Procedia Engineering 25:187–190
25. Yang J, Chen J, Liu Y, Yang W, Su Y, Wang ZL (2014) Triboelectrification-based organic film nanogenerator for acoustic energy harvesting and self-powered active acoustic sensing. ACS Nano 8(3):2649–2657
26. Shen S, Chiuand CY, Murch RD (2016) A dual-port triple-band L-probe microstrip patch rectenna for ambient RF energy harvesting. IEEE Antennas and Wireless Propagation Letters 16:3071–3074
27. Koohestani M, Tissier J, Latrach M (2020) A miniaturized printed rectenna for wireless RF energy harvesting around 2.45 GHz. AEU Int J Electron Commun 127:153478
28. Alex-Amor A, Palomares-Caballero A, Fernández-González JM, Padilla P, Marcos D, Sierra-Castañer M, Esteban J (2019) RF energy harvesting system based on an archimedean spiral antenna for low-power sensor applications. Sensors 19(6):1318
29. Lee J, Choi B (2014) Development of a piezoelectric energy harvesting system for implementing wireless sensors on the tires. Energy Convers Manag 78:32–38
30. Yoon Y-J, Park W-T, Li KHH, Ng YQ, Song Y (2013) A study of piezoelectric harvesters for low-level vibrations in wireless sensor networks. Int J Precis Eng Manuf 14(7):1257–1262
31. Yang J, Lee M, Park M-J, Jung S-Y, Kim J (2015) A 2.5-V, 160–μJ-output piezoelectric energy harvester and power management IC for batteryless wireless switch (BWS) applications. Proc Symp VLSI Circuits (VLSI Circuits):C282–C283
32. Wang W, Vinco A, Pavlov N, Wang N, Hayes M, O'Mathuna C (2013) A rotating machine acoustic emission monitoring system powered by multi-source energy harvester. In: Proc. 1st Int. workshop energy neutral Sens. Syst. (ENSSys), New York, NY, USA, pp 5:1–5:6

33. Sari I, Balkan T, Külah H (2010) An electromagnetic micro power generator for low-frequency environmental vibrations based on the frequency upconversion technique. J Microelectromech Syst 19(1):14–27
34. Gaur A, Tiwari S, Kumar C, Maiti P (2020) Polymer biowaste hybrid for enhanced piezoelectric energy harvesting. ACS Applied Electronic Materials 2:1426 –1432
35. Yu H, Yue Q (2016) Indoor light energy harvesting system for energy-aware wireless sensor node. Energy Procedia 16:1027 –1032
36. Iyer V, Bayati E, Nandakumar R, Majumdar A, Gollakota S (2018) Charging a smartphone across a room using lasers. Proceed ACM Interactive Mobile Wearable Ubiquitous Technol 1(4):1–21
37. Jabbar H, Jeong T (2022) Ambient light energy harvesting and numerical modeling of non-linear phenomena. Appl Sci 12(4):2068
38. Ahn D, Choi K (2018) Performance evaluation of thermoelectric energy harvesting system on operating rolling stock. Micromachines 9(7):359
39. Becker T, Kluge M, Schalk J et al (2009) Autonomous sensor nodes for aircraft structural health monitoring. IEEE Sensors J 9(11):1589–1595
40. Gao M, Su C, Cong J et al (2019) Harvesting thermoelectric energy from railway track. Energy 180:315 –329

Index

A. Nella et al. (eds.), *Energy Harvesting Trends for Low Power Compact Electronic Devices*, EAI/Springer Innovations in Communication and Computing,
https://doi.org/10.1007/978-3-031-35965-1